OUTLINES OF THE HISTORY OF BOTANY

OUTLINES OF THE HISTORY OF BOTANY

By

R. J. HARVEY- GIBSON

C.B.E., D.L., M.A.

GYAN

Published by

Gyan Publishing House
5, Ansari Road
Daryaganj, New Delhi-110002
Phone: 011-47034999, 9811692060
E-mail: books@gyanbooks.com

Distribution Network
gyanbooks.com
India, USA, Canada, UK, Australia, France

ISBN : 978-81-212-6571-3 (PB)
First Published, 1919

2nd Impression 2022

Printed at: Gyan Press, Delhi.

OUTLINES OF THE HISTORY OF BOTANY
Author: R. J. HARVEY- GIBSON

OUTLINES

OF THE

HISTORY OF BOTANY

BY

R. J. HARVEY-GIBSON

C.B.E., D.L., M.A.

FELLOW OF THE ROYAL SOCIETY OF EDINBURGH

MEMBER OF THE ROYAL DUBLIN SOCIETY

HOLBROOK GASKELL PROFESSOR OF BOTANY IN THE UNIVERSITY OF LIVERPOOL

1919

INSCRIBED BY PERMISSION

TO

H.E. THE RIGHT HON. THE EARL OF DERBY

K.G., G.C.V.O., LL.D.

H.M. AMBASSADOR TO THE FRENCH REPUBLIC

CHANCELLOR OF THE UNIVERSITY OF LIVERPOOL

PREFACE

THE following brief sketch of the history of botany makes no claim to originality save in the mode of presentation of the subject-matter. It comprises the substance of a course of lectures given to students of the University of Liverpool during their third year of residence, and has for its object the discussion of the more important features in the advance of botanical knowledge from the earliest times down to, approximately, the present day.

The ordinary university student, in the course of an already overcrowded curriculum, has only the minimum of time to devote to the study of the history of the science, and not always the knowledge or experience requisite to form a proper estimate of the relative values of the results achieved by the numberless investigators whose names appear on the pages of the larger text-books, still less to picture for himself in correct perspective, the evolution of the science as a whole. These lectures, it is hoped, may in some measure aid him in attaining these ends, so far as the history of botany is concerned.

Much has been omitted that another writer might have included; exigencies of time and space must serve as excuse for such omissions. Some aspects of the subject have been expanded, not unduly it is hoped, that others might prefer to treat more briefly or to omit altogether; that must be put down to individual idiosyncrasy on the part of the author. The style of the lecture-room has been retained in the hope that that mode of presentation may render the story more vivid and more readable.

The author desires to express his indebtedness to several friends who have given him the benefit of their criticism and advice. His thanks are especially due to Dr. D. H. Scott, F.R.S., and to Professor W. A. Herdman, F.R.S., who have very kindly read the whole work in manuscript, or in proof, and favoured him with many valuable suggestions. He desires also to record his acknowledgements to Professor F. O. Bower, F.R.S., and to Professor H. H. Dixon, F.R.S., for their kindly criticism of those sections on which they are recognised authorities. To his friend Miss M. Dunt he owes a debt of gratitude for constant help during the preparation of the text for publication.

HARTLEY BOTANICAL LABORATORIES,
UNIVERSITY OF LIVERPOOL,
1919.

CONTENTS

OUTLINES OF THE HISTORY OF BOTANY

LECTURE I

INTRODUCTORY

DURING the past two years, in the laboratory and in the field, you have been studying the structure and functions of plants, and, in the library and in the lecture room, you have become acquainted with some of the more theoretical and speculative aspects of the science of botany, as well as with the views of recognised authorities on the subjects you yourselves have been investigating. From time to time I have drawn your attention to the parts played by certain great men of past generations in the development of an idea, or in the discovery of some fact or law, jealously concealed by Nature from the uninquisitive or unsympathetic eye, and you have thus come to realise, more or less consciously, that the science itself has had a history, in the making of which many have co-operated, some largely, some modestly, but each contributing his fragment to the building of the whole. In this course of lectures I propose to sketch the story in a somewhat more connected form, to show you how the various aspects of botanical science first came to be appreciated, and how these aspects are related the one to the other. I hope also to present you with an estimate, of necessity a purely personal one, of the relative values of the labours carried out by the men of the past, and, indeed, though I trust with due deference, by the workers of the present also. In a word, I desire to conduct you to the top of a Mount Pisgah, whence you may survey the country through which you have travelled, and also view the promised land of future discovery and possible achievement.

But I would ask you to remember that what I propose to lay before you is merely an outline and not a detailed picture. When you stand on the top of a lofty mountain, a lake, a river, a village, a wood, may attract your attention in the plains below ; but a boat on the lake, a boulder on the river bank, children playing in the village street, or pine cones on the trees, are inconspicuous or altogether invisible. So when you take a bird's-eye view of the history of a great science your attention is necessarily directed in the first instance to the evolution of broad principles, to the works of the great masters, and trivial details, transient suggestions, and less prominent contributions of humbler labourers in the field are likely to remain, at most, blurred shadows on your mind, if indeed they make any impression on it at all. When you come to travel over the country you now view from the heights, it will be your task to study in fuller detail the more intimate features of the landscape now presented to you in broad outline only.

The Dawn of Botanical Knowledge

Not long ago I read, in a recently published text-book of botany, a sentence claiming that King Solomon, who flourished, we are told, some three thousand years ago, was the first professor of botany, because he " spake of trees, from the cedar tree that is in Lebanon, even unto the hyssop that springeth out of the wall." It would appear that his majesty was something of a zoologist also, for the ancient chronicler tells us that " he spake also of beasts and of fowl and of creeping things and of fishes."

It is much to be regretted that not one sentence of these early discourses on natural history has been preserved, unless we may assume that the fragments of ecological lore in the books of the Old Testament are some of the traditional sayings handed down by the "people who came to hear the wisdom of Solomon." "Can the rush grow up without mire? Can the flag grow without water," asked Bildad the Shuhite. "Whilst it is yet in his greenness and not cut down, it withereth before any other herb." "Wilt thou break a leaf driven to and fro?" suggests that Job had given some thought to the problems of stress and strain in mechanical tissues that formed the subject of Schwendener's investigations many centuries afterwards. And how does this strike you for a pronouncement in elementary plant physiology? "For there is hope of a tree, if it be cut down, that it will sprout again, and that the tender branch thereof will not cease. Though the root thereof wax old in the earth, and the stock thereof die in the ground; yet through the scent of water it will bud, and bring forth boughs like a plant." It might almost serve as the text to Alexander Braun's essay on *Rejuvenescence*, published in 1849!

But the study of practical botany must have begun long before the days of the learned King of Israel; for in the struggle for existence the problem of food supply is a crucial one, and primitive man was obliged, though perhaps unconsciously, to classify plants into those that were wholesome and those that were injurious. Not very long after the attainment of that elementary knowledge he would add to his stock of information his experiences of plants as stimulants, as sedatives or as curatives. In a word, our prehistoric forefathers must have studied plants both as sources of food and as sources of *materia medica*. What was good to eat and what was not was knowledge that each man had to gain for himself, perhaps from bitter experience, but the circumstances were altered when it came to the selection of the appropriate herb whose powdered root or leaf decoction might alleviate an internal malady or heal a wound. When his ailment was, as he generally believed it to be, inflicted on him by some supernatural being, it became necessary to consult a go-between, the medicine man, who, he supposed, had confidential dealings with the deity on the one hand and expert knowledge of the appropriate drugs on the other. The medicine man was thus a very important personage, more powerful often than the chieftain of the tribe himself. It is not necessary for you to tax your imagination in reconstructing the precise relations of the medicine man to the other members of the primitive human society; he is still in existence and still exercises his powers for good and for evil on his fellow tribesmen, as you may read in the pages of Sir J. G. Frazer's *Golden Bough.*

When we reach historical times the medicine man is replaced by the collector of *simplicia, i.e.* of the raw materials out of which the compound medicine is made, and by the physician who prescribes the drug. So, by and by, in the alleys of the old Greek cities there sprang into existence the shops of the sellers of drugs, or pharmacopolai, and of the collectors of roots, or rhizotomoi, while in the more select avenues the physicians, the real descendants of the medicine men, held consultations with their clients, diagnosed their ailments, and wrote the appropriate prescriptions.

But, as you will readily understand, something more was needed than a correct diagnosis of the complaint and a suitable prescription; the next essential was a pure drug, and the acquirement of that depended on the rhizotomoi who gathered, prepared, and sold, to the physician or to the patient, roots and leaves of medicinal repute. Probably many of the physicians were charlatans; it is certain that most of the apothecaries were. In Lucian's *Dialogues of the Gods* Hercules is made to call Aesculapius "a root digger and a wandering quack"; but some of the rhizotomoi whose names have come down to us had interests beyond the shop and the till, and experimented on themselves with new drugs they had discovered in the woods and the mountain glens. They tested the effects of vegetable poisons and found, what was rediscovered many centuries afterwards, that poisons might be given in small doses without any evil results, and that if the dose were gradually increased no injurious consequences followed, even when the amount, had it been administered in the first instance, would have proved fatal. "The virtues of all drugs," writes Theophrastus, "become weaker to those who are accustomed to them, and in some cases become entirely ineffective. . . . For it seems that some poisons become poisonous because they are unfamiliar, or perhaps it is a more accurate way of putting it to say that familiarity makes poisons non-poisonous; for when the constitution has accepted them and prevails over them, they cease to be poisons, as Thrasyas also remarked; for he said that the same thing was a poison to one and not to another; . . . also, besides the constitution, it is plain that use has something to do with it."

You must bear in mind that I am speaking of a period many hundreds of years before the era of the printed book, and hence all that the apothecaries of those days had to depend on for their knowledge of the correct herbs from which to prepare their "simples" was the laboriously copied manuscripts of their predecessors or the oral descriptions handed down from father to son or from master to apprentice. What wonder then that the physician often times complained of the inaccuracy of the apothecary's "make up," and blamed him for the non-success of his treatment of his patron's disease. Clearly the one thing wanting was an authoritative pronouncement of some kind on the morphological nature of plants, and a description of their constituent organs in precise terminology, that would form a means of accurately diagnosing plants of medicinal value. Before the herbalist and the physician could be sure of their ground a botanist had to be created. He was born in the person of Theophrastus of Eresus, one of the most distinguished botanists of all times.

For nearly a generation the English student has been brought up in the faith that the only *History of Botany* worthy of serious study is that by Julius von Sachs. That Sachs was a very distinguished botanist I am quite ready to concede; that he was always a fair-minded and discriminating historian of his science I take leave to doubt. In his estimate of Theophrastus he is singularly unfortunate, as you may now judge for yourselves by comparing the clear, simple, and, for the period when they were written, accurate descriptions of plants and plant organs in the translation of the *Historia Plantarum* by Sir Arthur Hort, with the dull crabbed phraseology of the German herbalists of the sixteenth century, whom Sachs would have us regard as the founders of modern botany. The truth would appear to be that Sachs had never read Theophrastus; he dismisses him in a couple of sentences on the assumption, as Lee Greene cynically suggests, that no one else would ever read him either.

Theophrastus was born in Lesbos, 370 B.C., and was a pupil of Plato and Aristotle. He does not appear to have travelled much, but to have derived his knowledge of plants chiefly from his daily observation of them in the garden, near Athens, which he had inherited from his friend and master Aristotle, in which he taught his pupils, and in which, when his long life at last came to an end, he was laid to rest.

The *Historia Plantarum* and the *De Causis Plantarum* are not long books, but at a time when all literature had to be laboriously transcribed by hand authors had to cultivate the virtue of brevity. Still Theophrastus manages to pack into his pages an astonishing amount of information; the two volumes were indeed veritable treasure-houses of facts for those who came after him.

He begins with an enumeration and definition of the parts of a plant—root, stem, branch, twig, leaf, flower, and fruit, and describes the first four as permanent organs, while leaf, flower, and fruit are classed as temporary or transient. The fruit, to Theophrastus, is comparable with the foetus of the animal, something produced by it but not a part of it; the ovary was merely the first beginnings of the fruit. That plants as a group had sex like animals he did not realise, nor, for that matter, did any other botanist for two thousand years after his day. For although he talks of trees as "male" and "female," "male" stands to him as a synonym for "barren." Thus he writes: "Taking, as was said, all trees according to their kinds, we find a number of differences. Common to them all is that by which men distinguish the 'male' and the 'female,' the latter bearing fruit, the former barren in some kinds." Dates and figs, however, seem to have puzzled him not a little, for in speaking of these plants he distinctly suggests sexual fusion. He writes: "With dates it is helpful to bring the male to the female; for it is the male which causes the fruit to persist and ripen, and this process some call, by analogy, 'the use of the wild fruit' (referring to the practice of caprification where the wild fig was employed). The process is thus performed; when the male palm is in flower, they at once cut off the spathe on which the flower is, just as it is, and shake the bloom with the flower and the dust over the fruit of the female, and, if this be done to it, it retains the fruit and does not shed it. In the case both of the fig and of the date it appears that the male renders aid to the female—for the fruit-bearing tree is called 'female'—but while in the latter case there is a union of the two sexes, in the former the result is brought about somewhat differently."

Plants, according to Theophrastus, are woody or herbaceous, and the herbaceous forms are classified as annual, biennial, and perennial. His garden apparently contained several hundred cultivated herbs, shrubs, and trees, at least he mentions some five hundred species in his two books. He recognised that the aerial supports of an ivy were roots and not tendrils, and that all underground parts are not necessarily roots; in other words, he appreciated, to some extent at least, the difference between true roots and rhizomes. The sepals and petals of the flower are modified leaves, so that the flower is to be regarded as a metamorphosed leafy branch. The root is defined as that part by which nutrient materials are taken up from the soil, a physiological definition, as is also his definition of the stem, viz. the chief means of transport of the nutrients to other parts. Theophrastus does not attempt to define a leaf, but he describes very fully the different kinds of leaves known to him.

Although he had no lenses, save those with which Nature had provided him, he shows himself possessed of a general knowledge of the rough anatomy of the different plant organs. He indicates the essential differences in stem, leaf, and seed between Monocotyledons and Dicotyledons, and has a fairly clear conception of the mode of formation of annual rings. Indeed he devotes two chapters of his *Enquiry* to anatomy, and nothing of any importance was added to his statements until the days of Grew and Malpighi in the seventeenth century A.D.

Ignorant as he was of the precise functions of sepals, petals, stamens, and carpels, he laid the foundations of our knowledge of floral morphology, for he distinguished between what he called "leafy flowered" and "capillary flowered" plants, *i.e.* between what we term petaloid and apetalous flowers, between hypogynous, perigynous, and epigynous insertion of floral leaves and "capillaries" or stamens, and between centripetal inflorescences like that of a foxglove and centrifugal ones like that of chickweed. He knew that a capitulum consisted of many flowers, a fact not always clearly realised even in the twentieth century by the junior student in our universities. The term "fruit" is applied to every structure that encloses seeds, and for the enclosing wall he coins the word "pericarp," a term still in use.

Many of you no doubt believe that the subscience Ecology was founded by Warming and Schimper only a few years ago, but you may read in the *Enquiry* accounts of woodland, marsh, lake, river, and other plant associations, with lists of plants most likely to be met with in each of these localities. The agriculturist and the horticulturist, as well as the pharmacist, will find in the *Enquiry* much that they ought to be acquainted with; thus a whole book is devoted to an account of the timbers of various trees and their uses. One curious notion was prevalent among the farmers of his day, viz. the belief that in certain circumstances one kind of grain could turn into another, but Theophrastus will have none of that. "Some say that wheat has been known to be produced from barley, and barley from wheat, or again both growing on the same stool; but these accounts should be taken as fabulous."

I hope I have now said enough to convince you that no one has a greater claim to the title of the "Father of botany" than Theophrastus of Eresus, in spite of the contemptuous treatment he receives at the hands of the German historian. After the famous Greek had been laid to rest among the plants he loved so well and studied so carefully, not a single pure botanist appears for more than eighteen centuries. There were, it is true, writers on agriculture like Cato, Varro, and Vergil, who flourished during the two centuries after the death of Theophrastus, and on materia medica like Dioscorides and Galen, but none of these men were really botanists as we understand the term. They dealt with the cultivation of plants as articles of food or as components of a pharmacopoeia, not as living organisms worthy of study for their own sakes.

Dioscorides, who flourished about A.D. 64, was a widely

travelled physician of Asia Minor, who compiled perhaps the most assiduously studied textbook that has ever been written. For over sixteen centuries its authority was supreme, and, as Lee Greene says, "the greater part of all the new botanical matter published during the whole of the sixteenth and part of the seventeenth centuries came out in the form of annotations upon the text of Dioscorides." Although his work is primarily pharmaceutical he recognises most of the more familiar natural families of plants, and, Lee Greene adds with justice and some indignation, "it is propagating fable in place of history to affirm that natural families were first recognised and indicated by any Linnaeus, or Adanson, or Jussieu of the eighteenth century," thus directly contradicting Sachs, who makes the assertion in his *History of Botany.*

Galen, who lived in the second century of the Christian era, was medical adviser to the Emperor Marcus Aurelius and also the author of a compendium of materia medica, in which he advocated that all physicians should acquire a knowledge of individual living plants, seeing that the average apothecary was not to be trusted to recognise them correctly.

The only other name that stands out prominently as a naturalist in these early years is that of Pliny, commonly called "the Elder," but his *Historia Naturalis* was little more that a popular compilation from Greek authors, a hotch-potch of fact and fable, without the slightest claim to rank with the works of Theophrastus and scarcely even with that of Dioscorides.

After Galen follows an absolute blank; for more than fourteen centuries botany had no history. Theophrastus had to be rediscovered, or rather all that he taught had to be relearnt, and, more important still, his spirit of enquiry had to be reborn, as we shall see it was in Valerius Cordus, a youthful lecturer in the Saxon University of Wittenberg. The story of this young genius's life is of great interest, and his contributions to the science of botany are recognised as fundamental by men like Tournefort and Haller, although he and his works together are dismissed by Sachs in less than a line as "of no importance."

In the Introduction to his *History*, Sachs tells us that "the first German composers of Herbals went straight to nature, described the wild plants growing round them and had figures of them carefully executed in wood." This is quite incorrect. Three of these sixteenth-century herbalists were Brunfels, Fuchs, and Bock. The first of them, by profession a Carthusian monk, expressly states that he described none of the plants he figures at first hand. "In this whole work," he writes, "I have no other end in view than that of giving a prop to fallen botany; to bring back to life a science almost extinct. And because this has seemed to me to be in no other way possible than by thrusting aside all the old herbals, and publishing new and really life-like engravings, and along with them *accurate descriptions extracted from ancient and trustworthy authors*, I have attempted both, using the greatest care and pains that both shall be faithfully done." The book is illustrated by about three hundred figures of variable degree of excellence; unfortunately, however, Brunfels, despite his protestations, was not always at pains to see that the description and the figure agreed. Lee Greene, who has made a special study of the texts of the sixteenth-century writers, tells us that "the Brunfelsian volumes are a treasury of select quotations from a long line of books, many of which are now seldom seen. But there are no new descriptions in his volumes; and it may be doubted whether upon the whole he advanced the art of plant description by a syllable." Brunfels's *Herbal* was published in Latin about the year 1530. He adopts the ancient classification of plants into woody and herbaceous, and rejects the alphabetical sequence of genera in favour of an association based on agreement in medicinal virtues. "The credit of having reformed the nomenclature of genera by the exclusion of names made up of two distinct words has been given to Linnaeus, who, in the year 1751, is thought first to have laid down such a principle. But the actual reform had been quietly inaugurated by Brunfels 220 years before Linnaeus came forward with his *Philosophia Botanica*. . . . To the nomenclature of species it is evident Brunfels gave no thought." He held "that for the science of botany there is an initial book; that is the *Historia Plantarum* of Theophrastus of Eresus," although he borrowed his descriptions chiefly from Italian authors who had attempted to identify the plants named by Dioscorides with those growing in the Mediterranean region.

Another ancient botanical worthy was Fuchs, whose *Historia Stirpium* appeared in 1542. In it the genera are arranged alphabetically and hence families are not recognised at all, and "there is no evidence that an interest in plants as plants rather than as drugs was ever awakened in him."

The third of the so-called "German Fathers" of botany was Bock, a contemporary of Fuchs and Brunfels; indeed it would appear that Bock wrote his herbal largely at the instigation of Brunfels. While Brunfels and Fuchs trusted to the plates with which their massive folios were embellished for the identification of the plants they described, Bock, who apparently was not overburdened with this world's goods, aimed at describing plants in such a manner as to be identifiable by the descriptions only, although woodcuts were afterwards added. His *Kreuterbuch* was published in 1546 and was in German, though afterwards translated into Latin. His method of classification is to associate "such plants as Nature seems to have linked together by similarity of form," *i.e.* his taxonomy is based on the vegetative organs only; for, though he seems to have studied the morphology of flowers and of fruits, he makes no use of them in arranging plants into groups. He had all sorts of weird ideas on biological subjects; for instance, he thought that orchids had no seeds but arose from the excreta of birds, while fungi were "merely the superfluous moisture of the earth and trees, of rotten wood, and other rotten things." Abiogenesis and the transmutation of one cereal into another, rejected by Theophrastus as "fabulous," are accepted by Bock as truths. He had one good point to his credit, however, viz. he described the wild plants of the countryside at first hand. He alone of the three paternal herbalists is entitled to the distinction of having gone direct to nature for his information.

"The tenor of the German writing of botanical history," says Lee Greene, "is that the science of botany was born again, as it were, in the year 1530 and in Germany, by the publication of Otho Brunfelsius' folio entitled *Herbarum Vivae Icones—Living Pictures of Herbs*. The Germans have always been and are the chief historians of botany. . . . All of them name Brunfels, Fuchs, and Tragus (Bock) as the fathers of the new botany of modern times. It has been indicated in a previous chapter of these *Landmarks* that the real father of botany as a science was Theophrastus of Eresus. If he is the father of the science he is the father of even modern botany, though not of those developments of it that have been the peculiar achievement of modern botanists. . . . We shall not be able to realise in how far the 'German Fathers' contributed to the superstructure of modern botany until we have examined with great care and diligence their best works; and this is something which, I shall make bold to say, not even the German historians have been at the pains of doing; though Sprengel, first of their lineage, did much and well in this direction, while also leaving very much for others to accomplish. Julius von Sachs, the last in the line, copied Sprengel's caption

The German Fathers, etc., but knew next to nothing of their works, even rating as unimportant Valerius Cordus, who was immeasurably the greatest of them all."

When we learn that Tournefort wrote of Cordus "in describendis plantis omnium primus excelluit," and that Ernst Meyer, from whom Sachs copied so largely, says: "Eine glänzende, nur zu flüchtige Erscheinung war des Euricius Cordus Sohn Valerius," concluding his short biographical sketch with the words: "So Vielfaches und Grosses in einem so kurzen Leben haben Wenige geleistet," it is difficult to understand how Sachs could have passed Cordus by with "several others of no importance," unless he simply could not be troubled to read him.

Valerius Cordus was born in 1515 of parents who, though poor, managed, like the traditional Scotch crofter, to save enough to send their son to the University. Valerius's father, himself a scholar of some repute, appears to have spared no pains to impart his learning to his son. Riffius, who contributed a preface to one of Cordus's books, writes thus of the youth: "To the best possible education of an intellect naturally keen there was united in him that happy temperament to which nothing is impossible or even difficult of attainment. To these gifts he added a truly marvellous industry and assiduity in research; and above all a most retentive memory for everything he either saw in nature or read in books."

At the University of Wittenberg Cordus became very intimate with a distinguished physician of Breslau named Crato von Kraftheim, who provided Conrad Gesner, ultimately the editor of Cordus's books, with a sketch of his life. Before he had reached his twentieth year he had published his *Dispensatorium*, practically a pharmacopoeia, while staying with an uncle who practised as an apothecary in Leipzig. Soon after taking his degree Cordus was appointed docent in the University, and in that capacity gave a course of lectures on Dioscorides, an epitome of which, called *Annotationes ad Dioscoridem*, appeared some years after his death.

His pre-eminence, however, does not rest on his *Dispensary* nor on his *Lecture Notes*; it rests on the fact that, as Haller puts it, he was "the first to teach men to cease from dependence on the poor descriptions of the ancients and to describe plants anew from Nature." During his residence at Wittenberg he appears to have explored the forests and wild mountain glens of the Erzgebirge, the Fichtelgebirge, and the Riesengebirge, and to have made several hundred new discoveries, surely no mean achievement for a youth scarcely out of his teens, with nothing to aid him in his search save a rehash of Dioscorides or Galen. As a result of these wanderings among his native mountains he wrote a *Historia Plantarum* when he was twenty-five years old, in which he described between four and five hundred species; but, in order to perfect his book before giving it to the printer, he went a pilgrimage through the Universities of Italy—Padua, Ferrara, Pisa, Bologna, Florence, Lucca, and finally Rome, where he fell ill of malaria and died in the 29th year of his age. His epitaph in the church of S. Maria de Anima includes the following lines:

> Ingenio superest Cordus: mens ipsa recepta est coelo:
> Quod terrae est, maxima Roma tenet.

The task of an editor of a posthumous work is never an easy one, but his duty would appear to be to try and visualise the form the publication would have taken had the author lived to guide its passage through the press, rather than to alter the original text or mode of presentation to bring them more into conformity with his own conceptions. Cordus's *Historia* was not published until 1561, seventeen years after his death, the MS. having been placed in the hands of Conrad Gesner. One of Cordus's chief aims had been to describe the plants he had found in such a way that plates might be dispensed with, but Gesner, most injudiciously, consented to the introduction of the crude woodcuts prepared for Bock's *Kreuterbuch*, with the result that new plants described for the first time by Cordus were confused with those discovered and figured by Bock. This is all the more unfortunate seeing that Tournefort, Linnaeus, and other later writers were thus led to believe that the illustrations had been prepared by Cordus himself.

The principles which Cordus followed in preparing his descriptions have fortunately been recorded for us by the author. He composed his account with the living plant in front of him, and he took care to see that the subject was full-grown and in flower, if not in fruit also. He begins his description with an account of the organs that first attract attention, the stem or leaves, and discusses the root last, as the least conspicuous. After stem and leaf he deals with the flower, giving at the same time the period of flowering. Then comes fruit and seed, where he mentions the number of loculi, the placentation, the mode of dehiscence in capsular fruits, and the form and colour of the seeds. The duration of herbaceous plants is always recorded. He adds data with regard to odour, flavour, and any economic values, such as the use of the plant in medicine. In a word, Cordus is primarily a botanist, and a medical man only in a very subordinate fashion. He had the courage also to describe afresh plants that had been outlined by his predecessors, when he felt that the accounts given by these authors were inaccurate, misleading, or imperfect.

Cordus's contributions to morphological botany were very considerable. He adopts Theophrastus's hint that not everything below ground is necessarily a root, and explicitly defines a rhizome as a "cauliculus" or little stem. Adventitious roots are recognised as such, and inflorescences receive the first scientific treatment they have had since the days of Theophrastus. Cordus was the first to recognise the bracts subtending inflorescences and speaks of them as forming an involucre. The calyx is distinguished from the corolla by position if not by name, and he unites ovary, stamens, and petals under one term—flower—for the sepals were not added till long afterwards. He appreciated clearly that the fig was an inflorescence, though Linnaeus, more than two centuries later, regarded the fig as a Cryptogam! He watched the periodic movements of leaves and the circumnutation of twining plants. He was the first to draw attention to the peculiar habits of the Sundew, and students of the physiology of nutrition will find in his pages the first mention of the tubercles on the roots of Leguminosae. How a fern multiplied had always been a puzzle to the ancients, but Cordus has no doubts about the process. "Trichomanes grows abundantly on moist shaded rocks, although it produces no stem or flower or seed. It reproduces itself by means of the dust that is developed on the backs of the leaves, as do all kinds of ferns; and let this statement of the fact once for all suffice."

Time will not permit of any further analysis of the achievements of this youthful botanical genius, but perhaps I have said enough to convince you that he is deserving of much more attention and commendation than is accorded to him in the chief history of botany available to you.

Summarising what we have learnt thus far, we see that scientific botany was founded three centuries B.C. by Theophrastus. Nearly four hundred years after him came the Greek physician Dioscorides, who regarded plants primarily as sources of materia medica, and one hundred years later still, Galen, who was also a pharmacist and even less of a botanist than Dioscorides. Then follows a gap of fourteen centuries till we meet with Brunfels and Fuchs, who did little more than rewrite Dioscorides and add more or less accurate pictorial representations to the text, and Bock, who, because he could not afford the expensive illustrations produced by his two wealthy contemporaries, tried to describe with greater care the wild flowers growing round him. Then came, lastly, the youthful genius on whom the mantle of the great Athenian had fallen, Valerius Cordus, and so the science that had slept for eighteen centuries reawakened to active life once more.

In the thirty or forty years during which Brunfels, Fuchs, Bock, and Cordus flourished we have only one name of any note to boast of in the history of English botany, namely that of William Turner. He was primarily a collector of plants and travelled extensively over Europe in quest of them. In 1551 and succeeding years he published a *History of Plants* of the usual herbal type, the plants being arranged alphabetically. This work was intended to replace the *Hortus Sanitatis*, a quaint production full of absurdities, that had long served as the text-book of medicine in England.

I have already mentioned to you the name of Conrad Gesner, the somewhat injudicious editor of Cordus's *Historia*. He is best known by a work of his own published in 1560, in which he treats much more fully of flowers and fruits than did his predecessors. He seems to have had some idea of what a genus meant, *i.e.* a natural group of species, and he certainly recognised the existence of varieties, for before acknowledging a species as such he insisted on proof of its constancy.

In the later years of the century two other treatises made their appearance, the *Rariorum Stirpium Historia* of Clusius, and the *Stirpium Adversaria Nova* of Lobelius. The former work was published in its final form in 1601 and gives descriptions of plants then regarded as of rare occurrence in Spain and in Austria-Hungary, while the latter production, published in 1576, was somewhat more ambitious and wider in its scope. Lobelius even makes an attempt at classifying his plants into groups, some of which, such as Cruciferae and Labiatae, are recognised to this day. The central feature of his classification is the use he makes of the leaf. He begins with long narrow-leafed forms like grasses, follows these with broader-leafed bulbous plants, and finishes with Dicotyledons, among which, however, he includes Ferns. He followed this principle also in laying out Lord Zouch's garden at Hackney, of which he had charge.

It is well that you should bear in mind the important part played by the garden in the early history of botany. The author of a herbal was, indeed, usually also the curator of a garden, sometimes a public one, but more often one owned by some wealthy noble or squire. Further, it was during the latter half of the sixteenth century that University botanic gardens began to be established and first of all in Italy. That of Padua was founded in 1545, of Pisa in 1547, and of Bologna in 1567; while later in the century other countries followed suit, at Leyden in the Netherlands, Montpellier in France, and Heidelberg in Germany.

Towards the close of the sixteenth century there flourished in the Netherlands a herbalist called Dodonaeus, whose work *Stirpium Historiae Pemptades Sex*, published in 1583, formed the basis of several of the English herbals; some indeed were frankly translations of the *Pemptades*, such as those of Lyte, and, to a large extent, of Gerard.

The only other outstanding name in the botany of this period is that of the Italian Caesalpino, whose *De Plantis* appeared in 1583. In this work botany as the study of plants is at last divorced from materia medica. It is interesting to note that the idea of separating the study of botany from medicine also occurred to contemporaries of Caesalpino. For instance, the little-known Bohemian writer Zaluzianski, in 1592, "pleads for the treatment of botany as a separate subject, and not as a mere branch of medicine." As this was a very important departure, I think it worth while to quote his own words as translated by Mrs. Arber in her interesting book on *Herbals*. "It is customary to connect

Medicine with Botany, yet scientific treatment demands that we should consider each separately. For the fact is that in every art, theory must be disconnected and separated from practice, and the two must be dealt with singly and individually in their proper order before they are united. And for that reason, in order that Botany (which is, as it were, a special branch of Physics) may form a unit by itself before it can be brought into connection with other sciences, it must be divided and unyoked from Medicine." The herbalist was content to arrange his plants in alphabetical order or according to their "vertues." Caesalpino, on the other hand, started with the idea that a system of classification could be established on certain theoretical principles, and, after meditating on these principles, he decided that the only natural system must be founded on the reproductive organs. The outcome was a series of groups in the highest degree unnatural, embracing plants which, as we now know, had not the slightest relationship to each other. In his methods Caesalpino thus showed himself an out-and-out Aristotelian. What Francis Bacon wrote of Aristotle equally applies to the Italian botanist. "For he had made up his mind beforehand; and did not consult experience in order to make right propositions and axioms, but when he had settled his system to his will, he twisted experience round, and made her bend to his system; so that in this way he is even more wrong than his modern followers, the Schoolmen, who have deserted experience altogether."

Some strange notions on physiology appear here and there in the *De Plantis*. Thus the chief function of leaves is to protect young buds and fruits from air and light; plants having no senses are unable to hunt for food, so they take it from the soil by suction; since plants need less food than animals they have no blood vessels, and so on. His entire work is tinctured with the strain of the Aristotelian philosophy, leading its author into all sorts of strange speculations, such for instance as to the precise localisation of the plant soul, which he finally decides must be just at the junction of the stem and the root.

It is really marvellous what an influence Aristotle and his philosophical conceptions exerted on the minds of these early explorers of nature, and indeed of thinkers in every branch of science and learning. A recent writer tells us that "Aristotle was almost as sacred as the Bible. At Oxford the University statutes enacted that (according to Giordano Bruno) Bachelors and Masters who did not follow Aristotle faithfully were liable to a fine of five shillings for every point of divergence and for every fault committed against the logic of the Organon!" Bruno lived in the latter half of the sixteenth century and was thus a contemporary of Caesalpino, so it is small wonder that Aristotelian ideas and methods should have made themselves evident in his works.

But the long night was passing and the dawn of a new day was at hand.

LECTURE II

INTRODUCTORY

THE reference in the last lecture to Caesalpino's crude ideas on physiology leads me to draw your attention to a matter of considerable importance, viz. the gradual evolution of the different aspects or departments of botanical knowledge. At first, as we have seen, botanical treatises were purely utilitarian—horticultural, agricultural, or medical, and the arrangements of the subject matter were devised primarily for convenience of reference. When plants began to be regarded as subjects of study for their own sakes and not merely as sources of drugs or as articles of food, an instinctive feeling was awakened that they must be related in some way to each other, in other words that they could be classified, but the problem was what was to be the basis of this classification. A division into herbs, shrubs, and trees was manifestly unsatisfactory, for the least observant botanist of the sixteenth century could not fail to notice that a tree like a laburnum had the same kind of flower and fruit as a pea, a pronounced herb. One way was to fix arbitrarily on some one organ as a distinguishing mark, and hence we meet with systems, such as those of Lobelius, where the criterion of affinity is the leaf, or of Caesalpino, where it is the seed and fruit. Some botanists, as we shall find later on, did not commit themselves to such artificial methods of classification, and yet seemed to content themselves with groping blindly after relationships, evidently hoping, like Micawber, that "something would turn up." That something did not turn up, however, for over three hundred years, when Darwin provided the key to phylogenetic relationship. In his *Philosophia Botanica* Linnaeus expressed the view quite definitely that plants could not be classified naturally by any "predetermined mark," but that it must be left to the future to discover the principles on which any natural taxonomy must be founded. When we come to deal with the work of Linnaeus we shall find that he attempted both methods.

The earlier classifications were naturally based on morphology only; the internal structure, or anatomy, was scarcely ever studied, for the simple reason that microscopes had not then been invented. Even when this instrument was added to the botanist's equipment it was long before microtechnique developed sufficiently to permit of the preparation of specimens that might reveal clearly the more minute structure of plant organs. Should you ever have occasion to visit the Royal Botanic Gardens in Edinburgh it will be quite worth your while to inspect the microscope used by Robert Brown in the beginning of the nineteenth century; I think when you have compared that curiosity with the elaborately finished instruments you are using in the laboratories to-day you will be led to wonder how so much was accomplished with appliances so primitive. Further, the "seeing eye" had to be developed and trained to interpret correctly what the microscope revealed. You yourselves must be conscious that, even with all the mechanical advantages now at your command, the power to express truthfully in a drawing what you see beneath a lens demands for its acquirement long and conscientious application.

Then again the lower plants formed a constant puzzle to the first botanists; they did not know what to make of them. These plants had apparently no seeds or other obvious means of reproducing themselves, although that neglected investigator, Cordus, you will remember, insisted that ferns grew from the " dust " produced on the backs of their leaves. Belief in abiogenesis was almost universal, and even as late as the end of the sixteenth century competent botanists thought that Fungi developed spontaneously from damp soil. Caesalpino, for instance, asserted that seedless plants were " bred of putrefaction." Ignorance of Cryptogamic Botany indeed was the chief cause of the long delay in discovering a unity of plan throughout the vegetable kingdom, a unity first brought to light by Hofmeister's epoch-making work half-way through the nineteenth century, although data had been accumulating for at least fifty years before his time.

From the earliest classical period plants were known to be living organisms, but this fact seemed to have been recognised only subconsciously, so to speak, for physiological investigations were not undertaken until well on in the eighteenth century. In this respect botany lagged far behind her sister science zoology. William Harvey, for instance, discovered the circulation of the blood in animals in 1628. Up to the sixteenth century all that was known about plant physiology could have almost been written on a sheet of notepaper. Naturally also the first observations were made as comparisons between plants and animals. The plant was a fixture and obtained its food supply from the soil, and that was practically all that was understood as to the physiology of nutrition; the part played by the leaf in relation to carbon assimilation was of course quite unknown. Had plants sex like animals? Theophrastus seems to have had a suspicion that at least some of them had, but his guarded hints were neglected, and even Caesalpino, though he founded his classificatory system on the fruit and seed, denied that plants had sex. Malpighi, towards the end of the seventeenth century, thought that stamens and floral leaves merely purified the sap that went to nourish the developing seeds. In spite of the fact that plants were fixtures, some of them were seen to exhibit movements of individual organs, such as the periodic foldings of the leaves of Leguminosae and the startling response to contact shown by the "Sensitive Plant." These observations were among the earliest beginnings of the section of plant physiology, sensitivity, that has attracted so much attention in recent years.

There is yet another department of botany to which I must draw your attention, but it is one of comparatively recent origin, and that is the study of fossil plants. It dates from the early years of the nineteenth century, but its importance was scarcely appreciated until more than half-way through it, until in fact phylogenetic classification demanded that the gaps in the system should, if possible, be filled up Palaeobotany is thus essentially a product of our own time.

Theophrastus was acquainted with about 500 plants; some 2000 are described in the sixteenth-century herbals, but at the present day the number has risen to something like 300,000. This tremendous increase has been due to the systematic search made by travellers in every quarter of the globe, and the subject of plant distribution or Geographical Botany now claims a volume to itself. Detailed examination of these finds soon made it evident that plants were not scattered haphazard over the earth's surface; that, on the contrary, their distribution was governed by a complex series of factors; that there were centres of distribution, barriers and aids to migration, and that the same plant type might appear in areas far apart from each other although nowhere between. Problems such as these linked themselves, on one side, to the discoveries of palaeobotany and, on the other, to plant physiology. In this latter relation a new line of investigation has made its debut quite recently, which you know under the name of Ecology, but which is after all merely the grosser physiology of plant communities.

You will thus realise that the field of botanical investigation soon grew far too vast to be cultivated by one individual. The *Historia Plantarum* of the early days, in which was discussed every aspect of plant life then studied, became broken up into treatises on morphology, anatomy, physiology, palaeobotany, ecology, and so on, and each of these is, in turn, a compendium of the detailed publications of workers who have devoted themselves to the investigation of, it may be, quite circumscribed areas of the field with which the subscience deals.

Before undertaking a survey of the work accomplished in the century following the era of Caesalpino I have thought this an appropriate opportunity to draw your attention to this subdivision of labour, so that you may be able to note the birth of a new line of investigation or the development of a new outlook. It is in the seventeenth century that we shall find that some of these new sections first make their appearance. As the years pass on the number of investigators increases by leaps and bounds, until at the present time their name is legion. It would not meet the end I have in view in preparing these lectures were I to attempt to mention even the names of all those who have contributed to the building of the *Theatrum Botanicum*, as the old herbalist, Parkinson, entitled his principal work. I propose to select comparatively very few authors, and if I omit many with whom you have at least a bowing acquaintance you will understand that it is not because I wish you to regard those I pass over as unimportant contributors to the sum total of our knowledge, but because I desire you to appreciate quite clearly, in the first instance, the principal landmarks in the history. You will be able to fill in the details for yourselves when you have time and opportunity, and when your own researches and more intensive reading compel you to master the results arrived at by other workers on the particular problems you may have set yourselves to solve.

The Foundations of Plant Morphology and Anatomy

A matter of forty years or thereabouts elapsed after the date of Caesalpino's treatise before the two brothers John and Kaspar Bauhin appeared on the botanical stage. The latter is by far the more important figure and his fame rests chiefly on two works which were published in 1620 and 1623. The earlier treatise was known as the *Prodromus Theatri Botanici* and the later as the *Pinax.* In the *Pinax* Bauhin quite definitely throws overboard the alphabetical arrangement of plants, and announces that any sound system of taxonomy must be founded on natural affinities. One of the first essentials was obviously to have a uniform method of naming plants, and to clear up once and for all the hopeless confusion caused by the careless use of synonyms. Plants in those days were provided with long descriptive sentences by way of names, much as if we were to employ the diagnoses in Hooker's Flora instead of saying simply *Ranunculus acris* or *Caltha palustris.* Bauhin was not perhaps always very successful in his efforts, but he certainly got the length of recognising the importance of using only two names to indicate the plant under consideration, and he did much to bring into general use the system of binomial nomenclature with which we are now so familiar. But if Bauhin distinguished genus and species and gave his plants christian names and surnames, so to speak, he made little or no progress in grouping genera into larger associations, or, as we would say, into orders and classes.

I must mention one other contemporary of Bauhin, viz. Jung, although his writings were not published till many years after his death. He made no attempt at founding a natural system, but devoted himself rather to working out such morphological problems as might serve as a basis for such a system. He has been described as a philosopher of the type of Caesalpino, but without Aristotelian obsessions, a close observer of nature as well as a keen thinker. To him both Ray and Linnaeus owe much, as we shall see when we come to consider the achievements of these men. Jung may be regarded as one of the first botanists of more recent times who studied morphology apart from taxonomy.

But morphology was not the only new department of botany to make its appearance in the middle of the seventeenth century. Lenses were coming into use as aids to natural eyesight. The more intimate structure of plant organs began to engage men's attention, and along with the knowledge thus acquired arose a desire to know what part the structures so revealed played in the plant economy. Plant anatomy and physiology were born, or rather reborn, but under vastly different conditions from those that prevailed when Theophrastus wrote his *Enquiry*.

Robert Hooke deserves our attention, not because he was a botanist (nowadays we would call him a microscopist) but because he was the first to employ the term "cell," and to recognise that vegetable tissues are composed of cells. You must not, however, run away with the idea that the word "cell" meant to Hooke what it means to you; far from it. Hooke had provided himself with a "magnifying glass" and promptly set to work to examine with its aid everything he could lay his hands on, and then he wrote a thick volume which he called *Micrographia*, or "some physiological descriptions of minute bodies, made by magnifying glasses, with observations and inquiries thereon." This tome appeared in 1665. It was during his microscopical examination of charcoal, cork, and other plant tissues that he first recognised that they were "all perforated and porous, much like a honeycomb." To these pores he gave the name of "cells," but the "interstitia," or cell walls, were not regarded by Hooke as constituent parts of the cells at all. I will quote to you one sentence only to show you that Hooke did not separate in his mind the anatomical structure from the physiological purpose. "Though I could not with my microscope, nor with my breath, nor yet any other way I have ever yet try'd, discover a passage out of any one of these cavities into another, yet I cannot thence conclude that therefore there are none such by which the *succus nutritius* or appropriate juices of vegetables may pass through them; for in several of these vegetables, whilst green, I have with my microscope plainly enough discover'd these cells fill'd with juices, and by degrees sweating them out." Observe how a preconceived idea governs Hooke's investigations and tempts him to prefer what he thinks must be the explanation to the evidence of his senses.

While Hooke, after the manner of the pure dilettante, was slicing up indiscriminately animal and plant tissues and studying their structure with the aid of his new toy, two men, Malpighi and Grew, the one in Italy and the other in England, were systematically examining and drawing vegetable tissues under the microscope, and laying the foundations of our knowledge of plant anatomy. Before their time, with the exception of Hooke's casual notes, there had been no attempt to make any advance on the meagre anatomical knowledge that had come down from the days of Theophrastus. In the hands of Malpighi and Grew this department of botany sprang into existence full fledged, and their monographs remained the standard works on the subject for well-nigh a century.

A considerable amount of controversy has taken place as to which of these two investigators has the claim to priority. The matter is of quite secondary importance, but the facts would appear to be as follows. Grew, who was a practising doctor, first in Coventry and afterwards in London, began work on plant anatomy in 1664 when he was 23 years of age, with the object of comparing plant and animal tissues. By the year 1670 he had accumulated sufficient material to justify him in writing out his results in the form of an essay which was brought to the notice of the then secretary of the recently founded (1645) Royal Society, and read before the Society in 1670. The MS. was printed in May 1671. Meanwhile Malpighi, who had been working independently on the same subject in Italy, had also written out his results and, strangely enough, sent his work also to the Royal Society, and, as it happened, an abstract of his MS. was read at a meeting of the Society in December 1671, when Grew's essay, now in print, was "laid on the table." Both works, therefore, bear the same date, although Grew is undoubtedly entitled to claim priority of publication. The second part of Grew's treatise appeared in May 1672, and a third in the spring of 1674, while the second part of Malpighi's work saw the light in August of the same year. The squabble between the partisans of these two pioneers was fortunately not reflected in the behaviour of the principals themselves, for each hailed the other's work with cordiality and appreciation. Indeed it is on record that Grew went so far as to propose discontinuing his researches in order to leave the field open for his rival, but, luckily for science, the Royal Society dissuaded him from such an exhibition of self-sacrificing renunciation.

I propose to sketch for you first of all Grew's main conclusions and then, much more briefly, Malpighi's views. I select Grew's work for detailed notice not merely on account of his prior claim so far as publication is concerned, but because his work is in English, while that of Malpighi is in Latin; you may thus consult Grew's volume for yourselves,—I regret the department does not possess a copy of Malpighi's essay. For your purposes it is really immaterial which of the two savants you study in order to gain some idea of the condition of anatomical knowledge in the closing years of the seventeenth century. There is one general point, however, to which I would draw your attention, and that is that in both treatises anatomical facts are closely interwoven with physiological theories. Indeed it would appear that the primary object of these investigations into structure was to discover how the machine carried out its duties. As Grew says in his preface, "By thus comparing of them (*i.e.* the several plants or parts of plants) we shall be able more exactly to state the orders and degrees of their affinities; better to understand both the causes and ends of their varieties and more probably to conjecture their natures and vertues."

The field of investigation was so new and so extensive and the means, looking at them from our standpoint, so inadequate that Grew is unable to refrain from making an anticipatory apology for his possible non-success. "I know it will be difficult," he says, "to make observations of this kind upon the organical parts of plants," but nevertheless he takes heart of grace in the following quaint words. "For what we obtain of Nature, we must not do it by commanding, but by courting of her . . . wherever men will go beyond phansie and imagination, depending upon the conduct of Divine wisdom, they must labour, hope, and persevere. And as the means propounded, are all necessary, so they may, in some measure, prove effectual. How far I promise not; the way is long and dark . . . the way of Nature, is so impervious, and, as I may say, downhill and uphill, that how far so ever we go, yet the surmounting of one difficulty, is wont still to give us the prospect of another. . . . To conclude, if but little should be effected, yet to design more, can do us no harm; for though a man shall never be able to hit stars by shooting at them; yet he shall come much nearer to them, than another that throws at apples." Consoling himself with these reflections, Grew begins his investigations.

His fundamental thesis is that every plant organ consists of two "organical parts essentially distinct," viz. the pithy part and the lignous part, or "such others as are analogous to either of these." In a seed "the pithy part" is composed of "parenchyma"—a term first introduced by Grew—and this is covered by a "cuticle." The parenchyma encloses an "inner body," the "lignous part," obviously the vascular system or at least the xylem part of it. He then describes, on the whole correctly, the chief stages in germination, but the underlying physiology is hopelessly wrong, because he had no conception of the existence and activities of protoplasm. "The general cause of the growth of a bean, or other seed, is fermentation. That is, the bean lying in the mould, and a moderate access of some moisture . . . being made, a gentle fermentation thence ariseth. By which, the bean swelling, and the sap still encreasing, the work thus proceeds." All through the writings of the early physiological anatomists, and indeed in those of their successors down to the beginning of the nineteenth century, you will find this idea of sap fermentation taking the place of what we nowadays regard as the functional activity of living protoplasm.

Grew next proceeds to examine the root and finds it to be composed of a skin, which he derives from the cuticle of the seed, and a cortical body "commonly called the barque" derived from the seed parenchyma. This cortical body is porous, the pores, *i.e.* the cells, being "innumerable" and "extream small." The central "lignous body" or vascular core, sometimes with and sometimes without a pith proper, is on the whole correctly described; and he shows also how the cortical body pierces "the lignous quite through as far as the pith." Obviously these "inserted pieces," as he calls them, are what we know as medullary rays.

How different was the physiological outlook in the days of Grew to that with which you are so familiar you may best realise from Grew's account of the functions of the different parts of the root, and which I may summarise in the following words: The root swells in the soil; some of the external fluids penetrate the cortex; this fluid is heterogeneous before it enters and becomes more so afterwards. In the cortex it ferments and the first result is the separation of a scum which passes outwards and becomes the skin. In its transit through the cortex the sap is filtered, at the same time the cortex is distended and new fluids enter. The filtered sap passes to the wood and has there added to it such extractives as the wood can provide. It now becomes "cambium" or refined plant juice, some of which is incorporated in the wood itself. By a kind of peristalsis the wood then pumps the remainder back to the cortex to be used in part as nutriment by that tissue; what is left over, "the second remainder," is discharged to the skin to nourish it in turn.

Grew next discusses the stem in the same way, recognising here also skin, cortical body, ligneous body, insertions, and pith. The ligneous body attracts his attention more especially; he describes the annual rings and likens the innumerable and "extraordinarily small vessels or concave fibres" of which it is composed, together with the "cortical insertions" or medullary rays, to the warp and woof of a piece of cloth. He searched in vain for valves in the vessels that might serve to direct the sap flow, but he presents us with the same strange notions about sap circulation in the stem that I have summarised apropos of the root. The upward movement of the sap is accounted for by capillarity aided by a force which suggests the pressure of turgid parenchyma, anticipating in a way the views of Godlewski and Westermaier two hundred years later. "And the said pipe or vessel," he says, "being all along surrounded by the like bladders, the sap therein, is still forced higher and higher; the bladders of the parenchyma being, as is said, so many cisterns of liquor, which transfuse their repeated supplies throughout the length of the pipe." He proved experimentally that the path of ascent was by the wood, but said that this path was followed in spring only, "for the greater part of the year, it riseth in the barque, *sc.* in the inner margin adjacent to the wood, and in the spring, in or through the wood itself, and there only."

Buds arise, he thinks, by an internal pressure exerted upon the wood by the turgid pith, the wood in turn pushing out the cortex and skin. He recognises the importance of the axillary origin of buds and the economy of space attained by the various methods of prefoliation, and makes the very astute remark that "a bulb is, as it were, a great bud under-ground." He is obviously puzzled to account for the endogenous origin of secondary roots, but he notes the difference in their mode of origin from that of buds.

The leaf has none of the fundamental functions that we are accustomed to ascribe to it, save that he recognises that water pumped up to it from the stem evaporates from its surface. Its principal use is the protection not only of its younger neighbours but also of the flowers and fruit, an idea he adopted from Caesalpino. Other duties of leaves are to remove the "grosser parts" or impurities of the sap and to act as reservoirs of superfluous materials. He appears to have identified the chloroplasts, for he talks of the air acting on "the acid and sulphurous parts of plants for the production of their verdure; that is, they strike all together into a green precipitate." It is air therefore and not light that induces greening, and this error is probably due to his idea that since alkalies can turn some plant extracts green, so some alkalies in the air can precipitate chloroplasts in the leaf. Grew has to his credit the first successful attempt to extract the green pigment from leaves, using oil as a solvent, and he further noticed that the solution was fluorescent. He also looked on stomata (illustrations of which he gives from the pine and the lily) as organs of transpiration, but he apparently thought that air was admitted by them also. "But as the skins of animals," he writes, "especially in some parts, are made with certain open pores or orifices, either for the reception, or the elimination of something for the benefit of the body; so likewise the skins of at least many plants are formed with several orifices or passports, either for the better avolation of superfluous sap, or for the admission of aer."

It is difficult to picture exactly what Grew thought was the part played by the atmosphere in plant physiology; " that even a plant lives partly on aer, for the reception whereof it hath those parts which are answerable to lungs" would suggest at first that he was confusing photosynthesis and respiration, but he goes on to tell us that this air is absorbed by the roots and thence distributed by the vessels and "insertions" to all parts of the plant's body.

Fifty years previously the philosopher Jung had made the unfortunate pronouncement "planta est corpus vivens, non sentiens," and Grew, like most of his successors indeed, accepted the aphorism as true and thus tried to explain the various movements of plant organs on purely physical grounds. When he found himself unable to put forward a mechanical explanation he solved the problem by referring the movement to an "inherent tendency." He entirely missed the conception of the plant as a sensitive organism responding to stimuli, but that idea was one that could not develop until the existence of protoplasm had been recognised and its peculiar activities appreciated.

Grew's conception of the more minute anatomy of the plant will seem very strange to you. "The most unfeigned and proper resemblance we can at present make of the whole body of a plant, is, to a piece of fine bone lace, when the women are working it upon the cushion; for the pith, insertions, and parenchyma of the barque, are all extream fine and perfect lace work; the fibres of the pith running horizontally, as do the threds in a piece of lace; and bounding the several bladders of the pith and barque, as the threds do the several holes of the lace; and making up the insertions without bladders, or with very small ones, as the same threds likewise do the close parts of the lace, which they call the clothwork. And lastly, both the lignous and aer vessels, stand all perpendicular, and so cross to the horizontal fibres of all the said parenchymatous parts; even as in a piece of lace upon the cushion, the pins do to the threds. The pins being also conceived to be tubular, and prolonged to any length; and the same lace work to be wrought many thousands of times over and over again, to any thickness or hight, according to the hight of any plant. And this is the true texture of a plant; and the general composure, not only of a branch, but of all other parts from the seed to the seed."

In view of the comparison of animal and plant skeletons I have had occasion to make to you in connection with the nature and distribution of mechanical tissues, it may be of interest to quote Grew's ideas on the subject. In his *Cosmologia Sacra*, published in 1701, he writes:— "In the woody parts of plants, which are their bones, the principles are so compounded, as to make them flexible without joynts, and also elastick. That so their roots may yield to stones, and their trunks to the wind, or other force, with a power of restitution. Whereas the bones of animals, being joynted, are made inflexible."

The parenchyma Grew likens to "froth of beer or eggs." "The microscope more precisely shows," he says, "that these pores are all, in a manner, spherical, in most plants; and this part, an infinite mass of little cells or bladders. The sides of none of these, are visibly pervious from one into another; but each is bounded within itself."

The structure of the vessels of the wood had a special fascination for the early anatomists. Grew's conception of a vessel was a ribbon wound spirally round a stick; the stick is then supposed to be withdrawn and the ribbon left in the form of a tube. The young wood, he believed, was formed from the inner layers of the rind, a view that continued to be held for many a day. Grew was a remarkably acute observer, and as evidence of this I may refer to his statement that in the spiral elements of the wood the spiral thickenings in the root run "by south, from west to east; but in the trunk, contrarily, by south, from east to west." In this observation he is perfectly correct.

I must now attempt to give you some idea of Grew's notions as to the structure and functions of the flower. You will remember that Theophrastus in his *Enquiry* mentions the fact that the date fruit does not mature unless the dust from the male flower be shaken over it, but it is doubtful if the Greek really believed that plants had sex like animals. Pliny reported the existence of two sexes in palms and talked of how the male palm "marries with other female palms, by gently sighing, tender looks, and the dispersion of a powder"! Among the botanists of the seventeenth century, however, the sexuality of plants was by no means an article of general belief. The English herbalist Parkinson, for instance, writes in his *Theatre of Plants,* published in 1640, "The ancient writers have set down many things of Dates—that they are male and female and that they both beare fruit so that they be within sight one of another, or else they will not beare—but I pray you account this among the rest of their fables."

Grew devotes a whole chapter to the structure of the flower, but the terminology he employs will appear somewhat bizarre to you. The calyx is called the "empalement," the corolla is the "foliature," and the stamens and possibly the styles also, the "attire." In a paper read to the Royal Society in 1676 Grew states that he had discussed the question of the functions of the floral organs with Sir Thomas Millington, at that time professor in Oxford, who had suggested to him that "the attire doth serve, as the male, for the generation of the seed," and that he, Grew, agreed with that view. The nearest that Grew gets to the idea of sex in flowers is his statement that the pollen "falls down upon the seed case or womb" and "touches it with a prolific virtue or subtle and vivific effluvia." Grew had noticed the visits of insects to flowers and observed how they carried away honey, wax, and "particular parts of the attire," but while recognising the benefit to the insect of these visits he missed entirely the advantages that accrue to the plant. The inward meaning of the give and take was not appreciated until the days of Conrad Sprengel, more than a century after Grew's time.

Having discussed Grew's work so fully it will not be necessary for me to spend much time over that of Malpighi. His conception of the anatomical structure of the stem resembles very closely that put forward by Grew. The outer part of the rind or cuticle, he thinks, consists of "utricles" laid down in horizontal rows, followed by layer after layer of "fibres," each fibre composed of tubes which open into each other. These layers form a network, the meshes being filled with roundish tubes. The wood which, like Grew, he thinks is formed by transformation of the inner rind, consists of fibres and tubes in longitudinal rows. Anastomosing spaces are left between the fibres and these are filled with bundles of tubes which run from the rind to the pith. These are obviously Grew's "inserted pieces" and our medullary rays. Among the bundles are "spiral tubes" in various situations but mostly in concentric circles; these are apparently the large tracheae of the spring wood, in which, strangely enough, he also managed to persuade himself that he had seen a kind of peristalsis taking place. In some parts Malpighi found other tubes containing milk, gum, turpentine, etc., and in others still he recognised minute utricles or bladders, which you are familiar with under the name of tyloses, but which he compared with the alveoli of the lung.

Malpighi gives the first published account of stomata, so far as I can find, for Grew merely refers to these organs as "orifices or passports." Malpighi says, "Among the vessels and network of fibres in most leaves are distributed special little air follicles or gaps which pour out either air or moisture. These gaps are especially evident in the leaves of the Oleander, where each ărea has four to many pores or mouths, the whole surrounded by a margin bearing numerous hairs," and on one of his plates the crypts and stomata, though crudely drawn, are indicated clearly enough. I am quite at a loss to understand how Sachs came to write that Malpighi "never succeeded in finding openings for the entrance of air in the roots or the leaves."

Here is Malpighi's conception of the functions of leaves. "Hence the active leaves seem to have been contrived by Nature for the digestion of food which is their chief function, for that part of the nutrient sap which enters the roots from below and which is not directed into the adjacent transverse branches, at length slowly reaches the leaves by their woody veins; this is necessary so that the sap should linger in the adjacent vesicles and so be mingled with the sap already there and be fermented. In this process the warmth of the surrounding atmosphere is of no little assistance, for it helps the more readily to evaporate that which is of no service. For this purpose Nature has provided the leaves with numerous special glands for the sweating forth and gradual elimination of moisture, so that the sap, being thereby condensed, may the more readily be digested in the leaves." From this you will see that Malpighi regarded stomata primarily as organs of transpiration.

Like Grew, Malpighi includes in his book an account of his investigations into the structure of the fruit, the seed, and embryo. His physiological conceptions as to the functions of the parts of the flower are quite as crude as those of Grew. Indeed he seems to have regarded the formation of seed as a kind of bud propagation, and thought that all the floral envelopes and stamens did was to remove certain fancied impurities from the sap that went to nourish the seed. Pollen, in short, is a sort of excrete product detrimental to good seed production.

Malpighi also gives a fairly correct account of the structure and germination of various seeds, such as the bean and the date. Like Cordus, he recognised the root tubercles of the Leguminosae, but of course made no attempt at determining their function.

LECTURE III

THE FOUNDERS OF NATURAL CLASSIFICATION

THE days of the herbalists with their catalogues of plants arranged in alphabetical order or in groups according to their medicinal and other "vertues" are now over, and at last plants are studied for their own sakes and not for what could be got out of them. A beginning had been made by Bauhin in establishing a uniform terminology, and morphology and anatomy had taught the botanists of the later years of the seventeenth century that it might be possible in time to evolve some sort of classification that would express the family relationships of the different genera and species to each other in a natural way.

One of the first to attempt such a classification was Robert Morison, whose chief work, the *Historia Plantarum Universalis Oxoniensis*, appeared in its complete form in 1680, when its author held the Chair of Botany in Oxford University. It would take far too much of our time to go into Morison's system in detail; moreover we shall find many of his ideas reappearing in the work of his greater contemporary Ray. Morison, strange to say, follows the old method of dividing plants into herbs, shrubs, and trees, subdividing these larger groups into sections based on the most varied features, some morphological, some physiological. Some of these sections contain plants of the most heterogeneous character; one, for instance, designated *Bacciferae*, includes Dicotyledons like Solanum, Sambucus, and Cyclamen, and also Monocotyledons like Arum, Paris, and Convallaria. "It seems almost incredible, but it is a fact," says Vines, "that the lapse of nearly 2000 years that separated Theophrastus from Morison marked no material advance in the science of classification." Morison appears to have been a rather cantankerous person and not over-just in his criticisms both of his predecessors, such as Bauhin and Caesalpino, and of his contemporaries, such as Ray. Perhaps the troublous times he lived in and the part he played in the Civil War—he was for a period a Royalist soldier—may have had something to do with the development of the pugnacious spirit so manifest in some of his writings.

A much finer character and a far more sympathetic student of nature was John Ray, the puritan divine, a man who has been described as "the greatest European botanist of the seventeenth century." While Oxford was the chief scene of Morison's labours, it was in the environs of Cambridge that Ray first studied the wild plants he met with in his rambles. From the Fenland he extended his wanderings into the Midlands, North Wales, Scotland, Cornwall, and the southern counties. After severing his connection with Cambridge owing to his finding himself unable to subscribe to the Act of Uniformity, he, along with his friend Willughby, travelled extensively over the Continent, returning after three years to complete his exploration of the homeland. He finally retired to a village in Essex where he spent the remainder of his life, writing up the results of his numerous journeyings and preparing his *Historia plantarum generalis*, described by Linnaeus and Haller as an "opus immensi laboris," and published in 1686.

Ray's principles of classification are a curious mixture of the views of Caesalpino and Jung, tinctured throughout with ideas of his own. He starts off with the old division of plants into herbs and trees; although he agrees with Caesalpino in regarding the fruit as important in determining the different groups, he seems to lay most weight on general vegetative habit; he appears also to have used anatomical characters in classification. He studied the structure of buds with care but failed to find them in herbaceous plants. The flower is made use of to a certain extent and it is quite obvious that Ray had some hazy idea that plants had sex. He examined the seed carefully and showed it to consist of an embryo and a "medulla" or "pulpa"—our endosperm. He used the number of cotyledons as a distinguishing mark between the two sections into which he divided his trees and herbs. To Ray therefore we owe the familiar names Monocotyledons and Dicotyledons as terms for these large groups of flowering plants.[1]

The *Historia*, in addition to formulating his scheme of taxonomy, gives a summary of morphology, anatomy, and physiology founded on Jung, Grew, and Malpighi. "Here may be found," says one of his biographers, "the principal discoveries in the nature of plants made by Caesalpino, Columna, Grew, Malpighi, and Jung, in addition to those made by Ray himself, and in this way resulted the most complete treatise which has yet appeared on vegetation in general." "The description of species," writes Sir John Smith, the first President of the Linnean Society, "is faithful and instructive, the remarks original, bounded only by the whole circuit of the botanical learning of that day; nor are generic characters neglected however vaguely they are assumed. Specific differences do not enter regularly into the author's plan, nor has he followed any uniform rules of nomenclature. So ample a transcript of the practical knowledge of such a botanist cannot but be a treasure."

Another work from Ray's pen, published in 1690, was the *Synopsis methodica stirpium Britannicarum*, a

[1] Ray was the first to define the word "species" as indicating the total of individuals which show constant characters from generation to generation.

small volume which may be regarded as the first British Flora. The book went through several editions and was for many a day the *vade mecum* of the naturalist on his botanical excursions.

Ray's works were well known both to the German and the French botanists. Rivinus, who was the leading Teutonic representative of the science, criticised Ray's system adversely and, in 1690, published a scheme of his own, in which the ancient division of the vegetable kingdom into herbs, shrubs, and trees was replaced by a classification founded on the corolla, an unfortunate selection, seeing that that floral character is perhaps the least reliable of all on which to base relationship.

Tournefort, in the closing years of the seventeenth century, was the leader in botanical thought in France, and, in his *Institutiones rei herbariae*, which appeared in 1700, formulated a thoroughly artificial system, also based on the corolla. It is somewhat remarkable that Tournefort, although his attention was centred so closely on the flower, did not appreciate as he might have done the revolutionary labours of Camerarius on the sexuality of plants any more than he did the views of Grew and Malpighi. Tournefort is often given the credit of having established and defined plant genera, but there does not appear to be much ground for this statement, seeing that Bauhin had already accomplished this eighty years previously.

The Sexuality of Plants

It is of greater interest to notice Ray's views on the sexuality of plants. The continental historians, and more especially Sachs, give the credit of the discovery of sex to Camerarius who wrote on the subject in a treatise called *De sexu plantarum*, in 1694. In justice to English botanists it may be worth while recalling what I have already told you of Grew's work. Ray in his *Historia* of 1686 says that "our countryman Grow supposes the stamens to perform the office of male, and that the pollen or little globules with which their apices are filled, and which separate from them when mature, serves the purpose of fructifying the parts which must be fecundated, and that the majority of plants are bisexual, that is, contain both sexes in the same corolla . . . nor is there occasion that the farina (pollen) should pass into the uterus (ovary) or the seeds, but only a subtle effluvia which is capable by itself of vivifying the included embryos." You will note that Ray interprets Grew as believing that the pollen does not actually induce the formation of an embryo but only stimulates it to further development. "This opinion of Grew, however," Ray continues, "of the use of the pollen before mentioned wants yet more decided proofs; we can only admit the doctrine as extremely probable." In his *Synopsis*, also published before the *De sexu plantarum* of Camerarius, Ray says, "Hence indeed is confirmed the opinion of those who teach that the dust contained in the apices of the stamens performs the functions of the male." As this was written four years before Camerarius wrote his book it is scarcely correct to say that the latter author actually *discovered* sexuality in plants. His contribution —a very important one without doubt—to the solution of the problem is really an experimental proof of the correctness of the theory suggested by Grew and supported by Ray, and to that extent he is entitled to full credit. This experimental proof was obtained by removing the stamens from flowers before the pollen was shed, in such cases as Maize, Dog's Mercury, Castor-oil, etc., and finding that in such circumstances no seeds were produced. From his experiments Camerarius was led to announce quite definitely that the stamens were male organs and that the ovary and style were female organs.

During the years immediately following the establishment of the doctrine of sexuality in the higher plants at least, the most contradictory opinions were expressed on the subject, some supporting the views of Ray and Camerarius, some denying them *in toto*. It is however quite unnecessary for you to trouble yourselves with these controversies, in view of the fact that opposition to the sexual theory gradually died out by the beginning of the eighteenth century.

The Physiology of Nutrition

The first fifty years of the eighteenth century were practically barren of research in plant anatomy; during all that period Grew and Malpighi remained the standard authorities. Some progress was made, however, in the physiology of nutrition. The physicist, Mariotte, who is probably well known to you as the discoverer of the law of gases that bears his name, wrote a treatise, published in 1717, long after his death, called *Sur le sujet des Plantes*, in which he combats the old Aristotelian belief that the chemical constituents of plants are obtained ready made from the soil. Mariotte, on the contrary, held that these proximate principles were manufactured in the plant body, and this view was, as we have seen, the one held by Malpighi, Grew, and others at the end of the preceding century. Naturally the questions that chiefly interested Mariotte were those concerned with the ascent of sap, which he attributed to capillarity, the pressure exerted by sap as evidenced by "spring bleeding," which he took to be the cause of the opening of buds and of the general expansion of the developing organs. In Mariotte's pages you will find the same attempt to discover homologies between the physiological processes in plants and in animals that we have recognised in all the earlier writers; the endosperm of the seed is the yolk of the egg; the latex corresponds to the arterial blood and the watery sap to venous blood, and so on. To Mariotte as to his successors, and also to many botanists up to the time of Ray, there was a special fascination about the circulation of sap, and efforts were constantly being made to draw a comparison between plant and animal in this relation. Ray, indeed, sought, but sought in vain, for valves that might regulate and direct the flow, and denied their existence only after he found he could induce water to pass down through the wood, as well as upwards and laterally.

Another physiological investigator of some note was Christian Wolff, who, in the early years of the eighteenth century, followed out some of the problems touched upon by Mariotte, Grew, and Malpighi. He studied the relation of the plant to the soil and observed that the latter was impoverished by repeated cultivation of crops upon it without periodic manuring. Rain, he thought, carried down substances into the soil, and the materials so added were absorbed by the root. While holding, with Mariotte, that capillarity was the chief cause of ascent of sap in the vessels of the wood he adds expansion of air as another factor in the process.

While Wolff may be regarded rather as a critic than as a discoverer, we must place a contemporary of his, Stephen Hales, in the very front rank as an experimenter. I have already pointed out to you how Grew, while laying the foundations of vegetable anatomy, had constantly in mind the solution of physiological problems, a task in which he cannot be said to have been very successful. Hales, on the other hand, tackled physiological questions, using as the foundation of his work Grew's anatomical results ; he is, in short, Grew's counterpart from the point of view of function, and the works of these two men taken together may be regarded as representing the state of plant anatomy and physiology in the first two or three decades of the eighteenth century.

Hales was born in the year in which Grew's *Anatomy* was published, viz., 1671, and died ninety years afterwards. For over forty years he lived the retired life of a country vicar, devoting all his leisure time to research both in botany and zoology. His numerous memoirs were published in a collected form in the volumes so well known to all students of botany as the *Statical Essays*, the first dealing with problems in plant physiology, the second with corresponding problems in animal physiology. The *Vegetable Staticks*, with which we are mainly concerned, is dated 1727, and in it we find evidence that entirely supports the title often given to its author of being the founder of the experimental physiology of plants.

You will remember the quaint ideas that Grew had on plant nutrition, the part played by various filtrations, fermentations, and what not. There is nothing of this vague theorising in Hales's book. The greater part of it is a record of a consecutive series of experiments, with a minimum of space devoted to deductions and general conclusions. Early in his career Hales had made experiments on arterial blood-pressure in animals, some of which would greatly shock the present-day anti-vivisectionist, and he expresses the wish that he "could have made the like experiments to discover the force of the sap in vegetables." An attempt to stop bleeding in a badly pruned vine, by means of a piece of bladder tied over the wound, gave him the idea of the manometer—"by mere accident I hit upon it," he says. Once again you will notice the research starts with an attempt to compare the functions of the animal and the plant; the arterial pressure in the one is to find its counterpart in the root pressure of the other. Having satisfactorily demonstrated this pressure he is naturally led to enquire as to its cause. He found that the pressure showed a daily periodicity, and that it was affected by changes in temperature. Since the pressure was upward only by way of the wood he saw that the term "circulation," in the sense in which it had been used by Harvey, was not applicable to the sap movements, and hence he denied the existence in plants of anything comparable with the arterial and venous flow in the animal. He noticed that leaves gave off water—"perspired"—or transpired as we say, and forthwith he proceeded to measure the amount transpired and to compare it with the amount absorbed by the root. He studied the variations in the quantity of water evaporated during the twenty-four hours and demonstrated the reduction in transpiration at night, and from a comparison of the amounts absorbed and transpired in a given time, he estimated the rate of ascent of the water-current. The forces concerned in the ascent were root pressure in the bleeding season and the "capillary sap vessels out of the bleeding season." "Though these have little power to protrude sap in any plenty beyond their orifices" (a comparison with the spurts of blood from a cut artery) "they can by their strong attraction (assisted by the genial warmth of the sun) supply the great quantities of sap drawn off by perspiration."

Having obtained successful results from one set of experiments, Hales at once uses these results as a basis for further enquiry. In this respect he exemplifies Grew's remark that "the generation of experiments is like that of discourse, where one thing introduceth an hundred more which otherwise would never have been thought of." One of these suggested investigations is into the connection between the plant and the atmosphere. You will remember that Malpighi was doubtful whether the air in plants enters by the root or by the leaf. Malpighi figures stomata, but is hazy as to their function; Grew, though he also knew of the existence of stomata, is equally vague as to their use to the plant. Hales, on the other hand, has some quite definite notions as to the part leaves play in plant nutrition. Transpiration is of course their chief function, but they have other duties as well. "For the air is full of acid and sulphureous particles which, constantly forming in the air, are doubtless very serviceable in promoting the work of vegetation; when being imbibed by the leaves, they may not improbably be the materials out of which the more subtle and refined principles of vegetables are formed . . . we may therefore reasonably conclude, that one great use of leaves is . . . to perform in some measure the same office for the support of the vegetable life, that the lungs of animals do for the support of the animal life; plants very probably drawing thro' their leaves some part of their nourishment from the air." That Hales studied the structure of the leaf for himself and was not content simply to accept Grew's word for the existence of stomata is manifest from his remark: "I found little or no air came either from the branches or leaves, except what lay in the furrows, and in the innumerable little pores of the leaves, which are plainly visible with the microscope."

The notion that "plants very probably draw through their leaves some part of their nourishment from the air" is a very shrewd guess indeed, for you must recollect that the composition of the air was not known until more than sixty years afterwards, and only very imperfectly even then. Leaves also absorbed light, according to Hales, which "may contribute much to the refining of the substances in the plant." Sachs in his account of Hales's work holds that he followed Newton in believing that light was actually a material substance. Hales indeed quotes Newton as asking, "Are not gross bodies and light convertible into one another? and may not bodies receive much of their activity from the particles of light, which enter their composition? The change of bodies into light, and of light into bodies, is very conformable to the course of nature, which seems delighted with transmutations."

Hales added nothing to our knowledge of the physiology of reproduction. "If I may be allowed to indulge conjecture," he begins, and then, very unlike himself, he follows on with an extremely quaint hypothesis in support of which he does not present any experimental proof. Here is his conception of the process of fertilisation almost in his own words: "Sulphur strongly attracts air; farina abounds in sulphur; it is placed in the movable apices of the stamina so as to be easily dispersed into the air, surrounding the plant as it were with a sublimed sulphureous pounce (powder) which uniting with the air particles may perhaps be inspired at several parts of the plant and especially at the pistillum and be thence conveyed to the capsula seminalis." Then follows a strange conceit: "and if to these united sulphureous and aereal particles we suppose some particles of light to be joined (for Sir Isaac Newton has found that sulphur attracts light strongly), then the result of these three by far the most active principles in nature, will be a *punctum* saliens to invigorate the seminal plant." You will note here a repetition of Grew's idea, the same sulphurous material and the same "vivific effluvia," but not a word about the actual fertilisation of the ovule.

"Hales," Sachs says, "is the last of the great naturalists who laid the foundations of vegetable physiology," but to my mind he is rather the first genuine plant physiologist we meet with in the history of the science. The very fact that he stands almost alone in founding his conclusions on actual experiments shows him to have possessed a truly scientific mind of the highest order. Some of his experiments, such as those concerned with the movements of sap, are quoted in modern textbooks of botany, and that surely points to Hales as a genius of an entirely different character to those who preceded him in this branch of botanical enquiry.

Linnaeus and the "Sexual" System of Classification

Just at the time when Hales was experimenting in his vicarage garden on the banks of the Thames, a young theological student was being initiated in the study of plants in the University of Lund in Sweden, who was destined to exert a tremendous influence on the science of botany in the generations to follow. His name was Carl Linnaeus. It is rather interesting to notice how many botanists of the past were born and bred in the atmosphere of the Church. Of those I have mentioned to you, Brunfels, Bock, Tournefort, Turner, Ray, Morison, Hales were all clerics or had at least embarked on a clerical career before turning to scientific studies. Linnaeus, as I have just said, was at first a theological student, and many of those I have yet to discuss with you were well-known divines or sons of such.

Linnaeus's life was an uneventful one. He was born in 1707, and when twenty-three years of age became curator of the Gardens of the University of Lund. The period from 1732 to 1738 he spent in travel in Lapland, Holland, England, and France, returning at length to Stockholm, where he practised medicine. In 1741 he was appointed to the chair of botany in the University of Upsala, where he remained until his death in 1778.

It is by no means easy to form a correct estimate of the effect of Linnaeus's work on botany. By some he is called the "Father" of the science, as if to indicate that he laid its foundations; by others he is looked upon as the last of the long line of systematists that began with Brunfels two hundred years before. Some have expressed the opinion that to Linnaeus the science owes an impetus that carried it on well into the nineteenth century, but there are not a few who think that his work was largely reactionary and did much to hinder the progress of botany. Let us try and see what he actually did accomplish and then come to some decision of our own on the matter.

I do not propose to enumerate all Linnaeus's contributions to botany, but only those that stand out prominently and in which his principal conclusions are formulated.

These are, I think, the *Systema Naturae*, 1735; the *Fundamenta Botanica*, 1736; the *Genera Plantarum*, 1737; the *Classes Plantarum*, 1738; and the *Philosophia Botanica*, 1751.

Linnaeus's outstanding characteristic is his power of describing and systematising. Sachs calls him a classifying, co-ordinating, and subordinating machine; he even classified the very botanists whose works he studied. He collected all the works of his predecessors, picked out all he thought best in them, and, with scissors and paste, welded the scraps into one concrete whole. Not that he palmed off the compilation as his own original production—far from it. Each author received due credit for what he had accomplished. He picked out a beam here, a tile there, a block of stone somewhere else, and out of the total mass of constructive material he built a house. His pre-eminent skill as a literary architect enabled him to erect a building that was hailed as a masterpiece both by his contemporaries and by generations of admiring pupils; a house perhaps convenient to dwell in at the time it was erected but unadaptable and quite unsuited to more modern requirements. It was as though the Hall of Cedric the Saxon had been offered to you as a residence in the twentieth century; where would you find accommodation for your modern furniture, even if you were successful in obtaining permission from sanitary, municipal, and other authorities to live in it at all?

Linnaeus himself was conscious that the scheme of classification he put forward was more or less of a makeshift, for in the *Philosophia Botanica*, written in the later years of his life, he attempted an entirely new system, which, however, he did not live to finish.

In trying to estimate Linnaeus's merits I cannot do better, to begin with, than quote the opinion of Bentham, his great successor in taxonomy: "It was reserved for the master mind of the immortal Swede to fix, by the establishment of genera and species, upon sound philosophical principles, a firm stage to serve as a basis and standing point for further progress and exploration. By his accurate discrimination of genera and species he really made possible the subsequent generalisations of De Jussieu and De Candolle." This statement expresses practically the unanimous view on Linnaeus's greatest achievement. He is credited also with the establishment of the binomial nomenclature, and with having, by the use of a terminology largely founded on, if not borrowed from, Jung, replaced the long-winded and confused descriptions of the herbalists by clear and succinct diagnoses. It is well to remember, however, that Linnaeus did not invent the binomial nomenclature; the germ of the idea is to be found, as you have already seen, in Bauhin's *Pinax*, and even in the *Enquiry* of Theophrastus.

Of anatomy Linnaeus knew nothing save what he learnt from Grew. Adopting the crude ideas of his predecessors he attempted to homologise the whorls of the flower with the successive concentric regions of the stem; the cortex becomes the calyx, the bast the corolla, the wood the stamens, and the pith the carpels. Sachs quite correctly says that Linnaeus showed an utter "incapacity for careful investigation of any object at all difficult to observe." In fact Linnaeus was not an investigator at all, for there is no evidence in his works that he made a single discovery of the slightest importance. Linnaeus does not go to Nature and invite her to tell him her secrets and then deduce from her answers what her purpose is, as Hales and Grew did; on the contrary he elaborated a complex and beautifully arranged and catalogued set of pigeon-holes and forced the facts that Nature presented to him into these pigeon-holes, whether they fitted the receptacles or not. In no aspect of his work is this utterly unscientific attitude of mind more clearly seen than in the famous artificial or "sexual system of classification" that was adopted and followed by students of botany for nearly a century afterwards.

In the *Philosophia Botanica,* it is true, Linnaeus does express the opinion that a satisfactory classification cannot be founded on pre-determined marks, but on natural affinities only. It is difficult for us in these post-Darwinian days to grasp precisely what Linnaeus meant by natural affinities. When we think of a natural classification we mean a genealogical one—a phylogenetic tree, in short—and by implication we assume that plants with similar characteristics have descended from a common stock. Linnaeus held no such views. In Sachs's words, "he assumed that plants of the highest and lowest groups of organisation were originally created at the same time and alongside of one another; no new class plants were afterwards created, but from the mingling together of the existing ones by the act of the Creator, generically distinct forms were produced, and the natural mingling of these give birth to species while varieties were mere chance variations from species." "There are just so many species," Linnaeus himself writes, "as in the beginning the Infinite Being created."

Linnaeus's chief merits, then, lie in his masterly powers of analysis, description, and diagnosis, and in his vigorous and effective advocacy of the binomial nomenclature, but his name is invariably and primarily associated with the so-called sexual system of classification. This title is quite misleading, for the system is not sexual at all in the true sense of the word. Linnaeus simply employed the numbers of the stamens and carpels, or, more accurately, styles as a convenient means of grouping plants together. It would be sheer waste of time to discuss the system in detail; you will find it set forth in nearly every textbook of systematic botany that treats of the historical aspect of the subject; but I must attempt in a sentence or two to indicate its fundamental features.

Linnaeus institutes 24 classes; the first ten of these

are based on the number of stamens only, 1, 2, 3, and so on up to 10; class 11 has 12 stamens; class 12 has more than 12 attached to the calyx, while class 13 has more than 12 attached to the receptacle. Then follow two classes, the first with didynamous, the second with tetradynamous stamens. Class 16 has all the stamens united by their filaments; class 17 has them united in two bundles; 18, in several bundles; 19, of plants having stamens united by their anthers; 20 has stamens and styles united; 21 and 22 have stamens and carpels in separate flowers, on the same or on different plants respectively; 23, which Linnaeus calls "Polygamia," has stamens and carpels separate in some flowers and united in others on the same or on two or three different plants. The 24th class, Cryptogamia, includes the entire host of lower forms from ferns and club mosses downwards, as well as some Phanerogams in which Linnaeus could find neither stamens nor carpels.

I need not, I feel sure, go further; if you really desire to appreciate the utterly retrograde nature of the system, all you need do is to take your Flora and write down under any one of these class headings the names of all the British plants you can find with the corresponding number of stamens, and then see for yourselves into what chaos the whole phylogenetic system is thrown. I shall be astonished if you do not agree with one recent writer who says that Linnaeus retarded progress in the attainment of a true natural classification of plants for nearly a century. How, then, can we account for the success of his system? In the first place it was delightfully simple and easy to work with, a sort of glorified city directory. Then again most botanical students of that period were inclined to hail with acclamation any scheme that saved them the trouble of thinking for themselves and deciding on the rival merits of the systems put forward by Morison, Ray, Rivinus, Tournefort, and others who were still groping for the true key to the problem. According to Linnaeus himself the highest and only worthy task of a botanist is to know all the species of the vegetable kingdom by name, and consequently the goal to be aimed at by every one who was ambitious of earning the title was to get to know the name of a plant in the easiest and most expeditious manner possible. If an unknown weed possessed five stamens, all of them the same length, it belonged to the class "Pentandria," and if, in addition, it had only one style, it was a member of the order "Monogynia" under Pentandria. It was not difficult thereafter to find the genus and species from the clear diagnoses that Linnaeus wrote in such a masterly manner. But a knowledge of the name thus acquired gave not the slightest hint as to its relationship to other plants with ten or any other number of stamens and five or any other number of styles, no more, in fact, than the knowledge that a certain individual is named Jones and pursues the occupation of a plumber in a street in Liverpool suggests that he is first cousin to a person of the name of Smith practising the same trade in a wynd in Edinburgh.

Linnaeus closes a chapter, he does not open one; he writes the epilogue, not the prologue; he summarises the past but sketches no outline of the future, save the unfinished fragments in the *Philosophia Botanica*. The chief service that he rendered to botany was to furnish his successors with a solid jumping-off place whence they might spring across the stream on to the further shore.

The Beginnings of Specialisation

We have now reached a period in the history of our science when developments begin to crowd upon us; the workers are increasing in number rapidly, and they are beginning more and more to specialise on definite problems, for the field is becoming too wide for any one man to compass.

The latter half of the eighteenth century saw little progress made in Taxonomy, for the Linnaean system was adopted, not only in Sweden but also in Germany and England, with an enthusiasm that spoke more for the reverence in which the authority of its inventor was held than for the scientific instinct of those who accepted and followed it. But in France the sexual system did not take a firm hold. In that country Ray's scheme was the one in greatest esteem, and this reputation was undoubtedly enhanced by the fact that it formed the basis of a new classification formulated by Antoine de Jussieu in 1789 in a treatise known as the *Genera Plantarum*. No doubt Antoine obtained many hints from his uncle, Bernard de Jussieu, who was the custodian of the Versailles gardens, where he laid out the plants according to a plan that was to all intents and purposes adapted from Ray, but of which he never gave any published account.

Antoine de Jussieu, like every other botanist of the period, was a firm believer in the constancy of species, and consequently we must not expect from him much more than a re-shuffling of the cards played by Ray and his contemporaries. He divided plants into Acotyledons, Monocotyledons, and Dicotyledons, extending the system published in the last edition of Ray's *Methodus*. The Acotyledons included plants like Naiadaceae as well as those we now call Cryptogams. The Monocotyledons are divided into three and the Dicotyledons into eleven classes, all based on the hypogynous, perigynous, or epigynous position of the stamens and petals in relation to the carpels, a most unfortunate selection of a distinguishing mark, as we now know. These fifteen classes are then subdivided into about one hundred "natural orders" or collections of genera, to which he assigned distinctive characters; this Sachs claims as De Jussieu's great merit.

In the department of morphology also the last fifty

years of the century were singularly barren. During that period there appeared only one work of any note, Gaertner's *De Fructibus et Seminibus Plantarum*, a laborious analysis of fruits and seeds, containing a large number of original observations, some of which were of great importance, though neglected for many years. For example, Gaertner insists that one-seeded indehiscent fruits are not seeds, a fact whose significance was not fully appreciated until the time of Robert Brown, forty years afterwards.

I do not intend to invite you to plunge into the metaphysical speculations of the poet Goethe on the subject of the metamorphosis of plant organs, where, as Sachs concisely puts it, he "confounds the subjective notion with the objective thing."

Goethe's theory of the morphological equivalence of appendicular organs was developed in his *Versuch, die Metamorphose der Pflanzen zu erklären* (1790), wherein he attempted to prove that there is one fundamental appendicular organ, the leaf, and that all other organs are modifications of it, and also that there is one fundamental plant type, the "Urpflanze," on the plan of which all others are built. A perusal of the book recalls the sophistries of Caesalpino and other Aristotelians, and prepares us for the sterile controversies that arose some years afterwards on the "spiral tendency in Nature" and on the precise limits of the "individual." Undoubtedly the "doctrine of metamorphosis" made a considerable impression on the minds of contemporary botanists, as also on those who lived in the later years of the century, but I think it is hardly worth while to spend time over a treatise so metaphysical. Indeed if we are to accept Schleiden's criticism we must regard Goethe's contributions to the science of botany as of extremely doubtful value. "The unfortunate seed which Goethe sowed," he says, "sprang up with sad rapidity; and next to Schellingism, we owe it to him that, in Botany, whims of the imagination have taken the place of earnest and acute investigation."

In anatomy Grew and Malpighi were still supreme, for Caspar Wolff's *Theoria Generationis*, published in 1759, can scarcely be considered as of primary importance. Like Goethe he concerned himself largely with the doctrine of metamorphosis and investigated the origin of foliar appendages at the "punctum vegetationis" of the stem. "In the entire plant," he says, "whose parts we wonder at as being, at the first glance, so extraordinarily diverse, I finally perceive, after mature consideration, and recognise nothing beyond leaves and stem (for the root may be regarded as a stem). Consequently all parts of the plant, except the stem, are modified leaves." You will note one marked difference between the methods of Goethe and of Wolff. The former arrived at his doctrine of metamorphosis by prolonged meditation on the appendicular organs of the vegetative and reproductive branches, while Wolff reached his conclusions by a close examination of the mode of development of these organs from primary meristem. Goethe's method was deductive, while Wolff's was inductive. It is said that when Goethe explained his doctrine to Schiller the latter commented, "this is not an observation, it is an idea."

In physiological research, on the other hand, the closing years of the century were particularly rich and well worthy of your earnest study and attention.

You will remember how, in 1686, Ray had admitted that the doctrine of sexuality in plants was "extremely probable" but required yet more decided proofs, and how Camerarius a few years afterwards provided some of these proofs by his experiments on castrated flowers. A further and important stage in the solution of the problem was reached when Koelreuter, about 1760, discovered the function of nectar and the part played by insects and also by wind in flower pollination. Koelreuter studied the structure of the pollen grain and identified the double wall and the sculpturing on the exterior, rather a feat when one considers the imperfect state of the microscope in his day. He noticed that something escaped from the pollen grain after it had lain for some time on the stigma, and fancied that this something was an oil which united with another oil secreted by the stigma, and that the combined fluids passed down the style to the ovary and there induced the formation of the embryo. In another direction also Koelreuter achieved considerable success. He tried the effect of artificially pollinating the stigma of one flower with the pollen of another species, and soon established the prepotency of the pollen taken from other flowers of the same species. He was successful also in obtaining hybrids between plants of different species, and thus in driving the first nail into the coffin of the dogma of the constancy of species.

The real connection, however, between floral morphology and insect visitation was discovered by Conrad Sprengel, who, in 1795, published his observations and deductions in a book whose title reflects the self-gratification of its author in the results he had achieved—*Das entdeckte Geheimnis der Natur im Bau und in der Befruchtung der Blumen*. Sprengel's main thesis is that the structure of the flower can be interpreted only by considering the duty of each part in relation to the visits of insects. The colour and scent are the sign-boards held out to attract the visitor's attention ; the markings on the corolla are guides showing the way to the hidden nectar supplies, like the indicative hands at our shop doors labelled " Jones's only entrance." Sprengel goes on to describe the various adaptations for the protection of nectar, and points out how hairs and scales shield it from rain, " just as a drop of sweat falling down a man's brow is stopped by the eyebrow and eyelash and hindered from running into the eye. . . . Insects can easily reach it but the rain cannot spoil it." Sprengel draws attention also to the order of ripening of stamens and carpels—dichogamy, and to the movements of the different parts of the flower and the significance of these movements in pollination. "Since very many flowers are dioecious," he says, "and probably at least as many hermaphrodite flowers are dichogamous, Nature appears not to have intended that any flower should be fertilised by its own pollen." There is the problem of cross pollination stated ; it was left for Charles Darwin to answer the question why this should be so. You will thus see that the labours of Koelreuter and Sprengel threw an entirely new light on the morphology of the flower, and placed in their appropriate niches all the isolated observations on the movements of stamens, of styles, of peduncles, and so forth that had puzzled the earlier naturalists.

LECTURE IV

THE PIONEER INVESTIGATORS IN PHOTOSYNTHESIS

MORE important even than the discoveries of Koelreuter and Sprengel in relation to pollination were those of Ingen-Housz, who, with the single exception of Hales, stands head and shoulders above all the plant physiologists of the eighteenth century. His work indeed lays the foundation-stone of all our knowledge of plant nutrition, and hence I must consider it in some detail.

In the year 1754 the Scottish chemist Black isolated and identified what he called "fixed air," rechristened "carbonic acid" some years later by Lavoisier. Mayow, in 1674, had already discovered a gas which he termed "spiritus nitroaereus" (oxygen), but the importance of his discovery was overlooked until, just one hundred years later, in 1774, Priestley rediscovered it and named it "dephlogisticated air," also renamed "oxygen" by Lavoisier. The publication of Priestley's *Experiments and Observations on Different Kinds of Air* marks an epoch in the history both of chemistry and of vegetable physiology. In this famous work the author speaks of "the restoration of air, in which a candle has burnt out, by vegetation," and puts forward the thesis that this purification is effected by "plants imbibing the phlogistic matter with which it is overloaded by the burning of inflammable bodies," and adds that he "generally found five or six days were sufficient to restore the air when the plant was in its vigour." Book IX. deals with "Observations and Experiments relating to Vegetation and Respiration," and in it he describes how a friend of his, Mr Walker (of English Dictionary fame), told him that, while waiting for a boat to convey him to the continent, he noted a horse trough at the principal inn at Harwich, which the landlord refused to have cleaned out because he found that its contents remained longer sweet when the sides and bottom of the trough were "covered by a green substance which is known to be of a vegetable nature." After describing the "spontaneous emission of dephlogisticated air from water containing a vegetative green matter" he says he "never found the emission took place save when the water was exposed to light."

Sachs in his estimate of Priestley's work says: "Priestley himself did not suspect that the deposit in question, afterwards known as 'Priestley's matter,' and found to consist of Algae, was a vegetable substance." This is quite incorrect. In Section VII. of Book IX. Priestley says that at first he supposed the green substance to be a plant but "was unable to discover the form of one. Several of my friends, however, better skilled in Botany than myself, never entertained any doubt of its being a plant; and I had afterwards the fullest conviction that it must be one. Mr. Bewley has lately observed the regular form of it by a microscope. It will come most properly under the denomination of the Conferva; but this not being within my province I shall not presume to give it any particular appellation." It would appear that Sachs had studied the works of Priestley with the same care that he bestowed on those of Theophrastus and Malpighi!

The seed sown by Priestley very soon bore fruit. In 1730, three years before Priestley first saw the light of day, there was born in Breda in Holland, Jean Ingen-Housz, who was destined to become one of the founders of plant physiology. After completing his university curriculum Ingen-Housz took up the practice of medicine, first in Holland and afterwards, about 1764, in England. Four years later he became physician to the Austrian Emperor and resided in Vienna. His tastes seem to have been scientific rather than medical, for during his life in the Austrian capital he began to send papers to the Royal Society of London, of which he was elected a Fellow on his return to England in 1778 or 1779. In the latter year Ingen-Housz published his *Experiments on Vegetables*, carried out in his garden near London, in leisure intervals of his professional labours among the dyspeptic attendants at the Court of George III. His discoveries are so important that I must quote them to you rather fully:

"The discovery of Dr Priestley," he writes, "that plants thrive better in foul air than in dephlogisticated air, and that plants have a power of correcting foul air, has thrown a new and important light upon the arrangements of this world. It shows, even to a demonstration, that the vegetable kingdom is subservient to the animal; and, vice versâ, that the air, spoiled and rendered noxious to animals by their breathing in it, serves to plants as a kind of nourishment." (Priestley speaks of "the pabulum which plants derive even from common air.") "When these observations are well considered, I think it will hardly be doubted but that there is something in the process of vegetation, or at least something usually attending it, that tends to ameliorate the air in which it is carried on, whatever be the proximate cause of this effect, whether it be the plants imbibing the phlogistic matter, as part of their nourishment, or whether the phlogiston unites with the vapour that is continually being exhaled from them; though of the two opinions I should incline to the former. The Rev. Dr. Priestley found that water, chiefly pump water, standing some days by itself formed at the bottom and sides of the vessel a kind of green matter, seemingly vegetable, from which air bubbles rise continually to the top of the jar, if exposed to sunshine, and that this air is fine dephlogisticated air."

Ingen-Housz then goes on to give the results of his own experiments and observations. These results are summarised in the preface to his volume, but as the statement is diffuse and couched in old-fashioned terminology it may be convenient to condense and translate it into rather more modern phraseology, using the terms with which we are now more familiar. "I observed that plants were able to purify bad air in a few hours if subjected to sunlight; that they could transform air absorbed from the exterior into oxygen; that this oxygen is exhaled into the atmosphere, thus rendering it more fit for animal life; that the exhalation of oxygen by plants begins after sunrise, is the more active the brighter the day and the more the plants are exposed to solar radiation, and the less active the more they are shaded by buildings or by other plants, when, so far from purifying the air, they contaminate it as animals do; that they cease to exhale oxygen as darkness comes on; that only leaves and petioles carry out this function; that the exhalation of oxygen in sunlight is independent of the poisonous or other quality of the plant; that the oxygen is given off chiefly by the under surfaces of leaves, and that mature leaves, *caeteris paribus*, give off more oxygen than young leaves; that some plants, especially aquatics, give off more oxygen than others; that, on the other hand, all plants in darkness render the air impure, especially such parts as flowers, fruits, and roots, no matter how economically useful they may be; that the sun alone has no power to purify the air—it may indeed tend to render it less pure—unless when acting in conjunction with green plants; and finally that the degree of purity of the air given off by green plants depends upon many factors such as the intensity of the light falling on them, the extent of exposure of the leaves to sunlight, and so on. Plants obtain most of their constituent juices from the soil by their roots and their carbon from the atmosphere whence they absorb air. This they elaborate in their leaves, separating from it the carbon they require for their own nourishment and throwing out the remainder which is useless to them but useful to animals, who in their turn take from the air, in the act of respiration, what they want and throw out the residue hurtful to them, but rendered serviceable once more to the plant. The oxygen yielded by the plant is elaborated by a kind of vital motion carried on in the leaf and kept up by the influence of sunlight."

From this summary I think you will at once recognise how great is the advance made by Ingen-Housz on the knowledge of the physiology of nutrition as described by Hales. No further progress was possible until chemists had established clearly the composition of the atmosphere, and Ingen-Housz is handicapped by the lack of this essential information. Still even in the *Experiments* he gives a fairly correct account of the gaseous interchange between the plant and the atmosphere. It is quite apparent, however, that to him the plant world is entirely subservient to the animal world and primarily a physiological apparatus for purifying the air to make it fit for animal use. He does not seem to realise that the whole of the animal world is dependent in the long run on the plant world for its nutriment.

In 1796 Ingen-Housz produced another work entitled *On the Nutrition of Plants and the Fruitfulness of the Earth*, in which he shows that he is now armed with the knowledge he had acquired of the new chemistry founded by Lavoisier, a knowledge he confesses he did not possess when he wrote the *Experiments*. He knew now that carbon dioxide was a compound of carbon and oxygen, and this at once enabled him to grasp the significance of the gaseous interchanges taking place between the green leaf and the air in sunlight. His new interpretation of the phenomenon is that the leafy shoots give off oxygen in light and carbon dioxide in darkness, while non-green parts give off carbon dioxide both in the light and in the dark. From the carbon and oxygen the plants construct "acids, oils, mucilage, etc.," and these bodies are then combined with the nitrogen of the air. This latter idea is of course erroneous, but it took another fifty years to prove it so. That, unfortunately, was not the only blunder he made. Although he admits that the leaves absorb carbon dioxide he still thinks that a considerable proportion is obtained by the roots from the soil. He also falls into the error of believing that the plant gets its oxygen from the carbon dioxide by night and in the shade, and carbon from the same source in sunlight, giving off the oxygen and retaining the carbon as a nutrient. I cannot find in Ingen-Housz's pages any justification for Sachs' statement that he "not only discovered the assimilation of carbon and true respiration of plants but also kept the conditions and the meaning of the two phenomena distinct from one another." But even if we do not credit him with all that Sachs claims for him, we must admit that Ingen-Housz is fully worthy of a place in the front rank of those who laid the foundations of our knowledge of plant nutrition.

A contemporary of Ingen-Housz was the Swiss botanist Senebier, who in the eight years between 1782 and 1790 published a series of treatises on the subject of plant nutrition. The *Mémoires* appeared in 1782; the *Recherches* in 1783; the *Expériences* in 1788, and the *Physiologie végétale* in 1790. Senebier recognised that leaves are essential organs to all plants which possess them, but he expressed astonishment that some plants have no leaves and yet, having green stems, give off oxygen in sunlight, thus missing the important generalisation that function need not be dependent on morphological value. Many of Senebier's conclusions are merely restatements of those of Ingen-Housz, with whom he keeps up a constant polemic. Thus he finds that the quantity of gas given off by a plant cannot be accounted for by the amount present in the plant body, and takes great pains to prove that the gas evolved from green parts differs entirely from that exhaled by roots and flowers. He only confirms his contemporary when he states that young leaves are unable to make full use of the carbon dioxide supplied to them. He always seems to miss the fundamental generalisation—" only green plant organs carry out carbon assimilation." He finds that leaves perform their functions when isolated as well as when attached to the plant. By a long series of experiments he shows that it is light and not heat that is the effective agent in the assimilatory process. He finds also that oxygen is given off only when carbon dioxide is present, but comes to the extraordinary conclusion that "fixed air" is changed into oxygen and that this change is brought about by its acid character acting as a stimulant on the leaf. Impressed with this idea he tried to produce the same results by using other acids, but needless to say the plants died. He said that aquatics give off the more oxygen the more the water in which they are cultivated is impregnated with "fixed air," a conclusion at variance with that arrived at by Ingen-Housz, who, after describing a similar experiment, says: "It is also true that leaves thus placed in water impregnated with 'fixed air,' do not yield that fine dephlogisticated air which they yield when placed in common pump water."

Although Senebier has a vague idea that the "fixed air" has an important relation to the vital economy of the plant he fails to grasp the fact that the oxygen comes from the carbon dioxide, for he thinks that the leaf in the sunlight can change nitrogen into oxygen and that "fixed air" may also be transformed into oxygen. In spite of very many experiments Senebier persists in regarding the soil as the source of the "fixed air" and the root as the absorbing organ. Finally, Senebier flatly contradicts Ingen-Housz in his views on gaseous exchange in darkness—"Non," he says, "les plantes ne produisent point de l'air fixe quand ellesne fermentent pas."

I cannot see on what grounds Pfeffer bases his statement (in his *Plant Physiology*) that Senebier "established the fact that organic substance is produced from carbonic acid gas and water, while oxygen is excreted." Senebier certainly did not "make it clear that the exhalation of oxygen was accompanied by a corresponding decomposition of carbon dioxide," although Pfeffer asserts that this discovery was "reserved for Senebier." On the contrary, I can find no evidence that leads me to believe that, as Hansen, a later historian of the subject, puts it, Senebier was either the discoverer or even the co-discoverer of carbon assimilation.

I have passed over the names of several men who figured in the history of plant physiology during the eighteenth century but previous to the days of Ingen-Housz and Senebier. In a sketch so brief as that I am giving you I should not be justified in devoting more than a sentence or two to their achievements. One of these investigators was De la Baisse, who wrote about the year 1735. He attempted to follow the path of ascent of sap by watering the plant with coloured fluids or by placing cut branches in them. Another worthy, called Reichel, also used coloured fluids and identified these in the lumina of the vessels and held, therefore, that vessels carry sap and not air, "an unfounded notion," according to Sachs, and one which Sachs himself attempted to disprove 150 years afterwards, but without success.

Yet another contribution—if contribution it may be called—came from the pen of Charles Bonnet, whose pretentious work on the functions of leaves appeared in 1754. The thesis he sets out to prove is one he adopted as a "sensible suggestion" from the mathematician

Calandrini, of Geneva, viz. that the under side of the leaf absorbed the dew that rose (!) from the soil. Bonnet's rank among the pioneers in plant physiology cannot be more tersely expressed than in Hansen's scathing words: "Whenever a history of false prophets in science comes to be written then Bonnet will come into his own, for he will be found marching in the very front rank."

The only other name I need mention is that of Du Hamel, who, in 1758, compiled an account taken from the works of Malpighi, Mariotte, Hales, Bonnet, and others. When I tell you that he thought the soil was a sort of digestive organ which prepared the food that the root absorbed, and that the leaves were suction pumps for drawing it up, I think you will admit that Du Hamel's contributions to the theory of nutrition are not likely to repay much expenditure of time on their study. In one direction, however, viz. the movements of plant organs, Du Hamel seems to have made some observations worth recording. Ray had noticed the periodic movements of leaves of Leguminosae and the heliotropic curvatures of shoots, and attributed them to changes in temperature. Dodart attempted to explain geotropic curvature of roots and apogeotropic curvature of stems by assuming the contractility of fibres on the damper side of the root and the drier side of the stem. Du Hamel, on the other hand, asserted correctly that heliotropic curvature was dependent on light, but that movements of Mimosa leaves were independent of it seeing that they took place in darkness also. Evidently he did not continue his experiments for a long enough period. He also extended Hales's pioneer observations on growth, confirming that acute observer's statement that growth in length is confined to the apical region of the root.

Although no connected research was carried out during the eighteenth century on plant movements in general, a considerable number of scattered observations had been recorded, such as the spontaneous or contact movements of stamens, the " circulation " of cell contents in some Thallophyta, the oscillation of the filaments of certain Schizophyceae, and so on. All these mysterious phenomena were explained either on purely physical grounds or, especially towards the end of the century, by dragging in a mysterious " vital force " which most conveniently relieved the enquirer of the trouble of making any further investigation. It was not indeed until the early decades of the nineteenth century that botanists began to feel dissatisfied with this explanation—which was in reality no explanation at all but merely a confession of ignorance of the true cause of such movements, viz. the sensitivity of living protoplasm to external stimuli. You will be better able to appreciate this change of outlook after we have considered the experimental researches of Thomas Andrew Knight on geotropism and heliotropism.

In a previous lecture I asked you to notice how the science of botany would be likely, as time went on, to progress along certain fairly well defined lines. As you doubtless remember, starting from the purely utilitarian cataloguing and describing of the earlier herbalists, there appeared, first of all a more or less conscious effort to group plants according to their natural affinities, to discover, in other words, a rational classification. The search for the principles on which such a classification could be based led as a matter of course to the investigation of the different organs of plants with the view of discovering which of them might form the most satisfactory and reliable " marks " to follow in arranging and grouping the members of the vegetable kingdom as a whole. The independent study of morphology led men to probe more deeply into the structure of plants, and hence arose the subject of anatomy, which is really morphology extended and intensified.

It is not consonant with the scientific mind to investigate the structure of a piece of mechanism and refrain from asking what it is for, what are its functions. If it be a living mechanism, obviously the first question that demands an answer is, how does it live? There were some men in the eighteenth century who attempted to answer this question, and to them we owe the foundation of our knowledge of the physiology of nutrition. But plants manifestly not only grew but multiplied, and an investigation of the organs of propagation led naturally to enquiries as to how they carried out their duties, to a physiology of reproduction. The acceptance of the totally misleading aphorism of Linnaeus, "minerals grow, plants grow and live, animals grow, live, and feel," prevented men from recognising as yet a physiology of sensitivity, while the other departments of the science to which I have directed your attention were not even dreamt of.

This seems an appropriate moment to pause and take stock, so to speak, of what had been accomplished during the eighteenth century. In taxonomy the most noticeable feature is the gradual development of the idea that a natural classification of plants was attainable if only the fundamental principles could be discovered, and the earlier years of the century present us with several tentative efforts in this direction, notably those of Morison, Ray, and Tournefort. But all these groupings, somewhat erratic perhaps, but still in the right direction, were suddenly eclipsed by the so-called "sexual system" of Linnaeus, which, though it may have proved a temporary convenience, did much to stifle all progress in real taxonomy for nearly a hundred years. It was fortunate that the lamp still burned, however feebly, in the hands of Antoine de Jussieu, just as the century was drawing to its close.

In morphology no great progress had been made save in relation to fruits and seeds. Men still subsisted on the fragments of classical lore that had filtered down through the herbalists, along with the somewhat meagre additions made by Jung, Ray, and Linnaeus. In anatomy however, thanks to the evolution of the microscope, matters were in a much more advanced state. The classic monographs of Malpighi and Grew stand out like lighthouses in the darkness, and there is nothing to replace them until long after the dawn of the nineteenth century.

Although, as you are well aware, we are not, strictly speaking, justified in talking of the development of the idea of sexuality in plants until the discovery of the actual gametes concerned in the process, still, accepting the interpretation of the phenomena as then understood, we may say that by the end of the eighteenth century it had come to be accepted that the stamens corresponded to male organs and the carpels to female organs, and that the accessory parts of the flower were intimately concerned in the process of pollination, or the application of the pollen to the stigma, which had come to be recognised as an essential preliminary to the formation of the seed. What happened inside the ovary and what part the pollen played in these happenings was as yet a mystery. In the department of nutritive physiology the hazy conceptions of Malpighi and of Grew, with their weird ideas on the intermingling of juices and subsequent fermentations, were beginning to give place to conclusions based on definite experiments. Hales stands out as the pioneer investigator on the new lines, and to him we owe the foundations of our knowledge of root pressure, transpiration, and the ascent of sap. But undoubtedly the greatest advances in the theory of plant nutrition were made after 1774 when Priestley rediscovered oxygen gas. Ingen-Housz overshadows all others in the closing years of the century as the discoverer of the significance of the gaseous interchanges taking place between the leaf and the atmosphere, and his work, taking into account the condition of chemistry in his day, must be regarded as one of the greatest contributions to our knowledge of plant physiology that has ever been made by any one man

Progress in Taxonomy and Anatomy at the Beginning of the Nineteenth Century

The next important step in advance in the natural classification of plants was taken by two representatives of a very distinguished family, that of De Candolle in Switzerland. August Pyrame de Candolle, the elder of the two men whose names have become household words in the history of botany, based his classification on that of De Jussieu, but modified it greatly in accordance with certain principles which he elaborated in a volume called *Théorie élémentaire de la botanique*, published in 1813. The chief thesis he sets out to establish is that morphology is the key to taxonomy, and that physiology for this purpose is not only useless but actually misleading. His first task is to define and illustrate the doctrine of symmetry of organs. This, he says, necessitates the comparison of a large number of forms, and such a comparison discloses the fact that the difficulties in the way of the recognition of relationship between forms that are in reality quite closely akin are due to three causes, abortion of parts, degeneration of parts, and adherence of parts of one kind to parts of another kind. "The whole art of natural classification," he says, "consists in discovering the plan of symmetry."

While holding such views, De Candolle was, strange to say, a believer in the doctrine of the constancy of species, and it is remarkable that he succeeded in maintaining this belief and at the same time propounding the view that organs might degenerate, or even abort, or unite with other organs. If a plant possessed four stamens and a staminode when, according to the doctrine of symmetry it ought to have had five functional stamens, it must either have been created with a staminode in place of the fifth functional stamen, and then there could be no degeneration, or it must have been derived from a plant with five functional stamens, and then there could be no constancy. There is no way out of the dilemma, and it is strange that De Candolle did not appreciate this. Did he also, like Goethe, confuse the "subjective notion" with the "objective thing"?

De Candolle put forward two classifications, the first in 1815 and the second in 1819. In the first of these attempts he adopts De Jussieu's plan of forming three main groups, Dicotyledons, Monocotyledons, and Acotyledons, but he introduces some new ideas into the subdivision of these. Thus the Dicotyledons are divided into sections based on the degree of fusion of the petals and their position in relation to the ovary. The first division includes plants which are polypetalous and hypogynous, while the second comprises those that are polypetalous and perigynous. On that system some of the Saxifragaceae, as you know them, would appear in the first division, some in the second, and some would be excluded from both. Again, the Monocotyledons are, according to De Candolle, "phaenogamous" where the reproductive organs are exposed and regular, and "cryptogamous" where they are concealed, irregular, or unknown. The Acotyledons are divided into leafy and sexual and leafless and asexual. De Candolle seems himself to have had no great conceit of his first attempt, for four years later he published a second scheme in which he departs from the fundamental principle he had originally laid down, dividing plants into vascular or cotyledonous and cellular or acotyledonous, subdividing the former into Exogens (Dicotyledons) and Endogens (Monocotyledons) on quite erroneous anatomical grounds. The Exogens again are divided as follows :

A. With calyx and corolla—Dichlamydeous.

1. Thalamiflorae—polypetalous and hypogynous.
2. Calyciflorae—polypetalous and perigynous.
3. Corolliflorae—gamopetalous.

B. With a single perianth—Monochlamydeous.

The Endogens are still "phanerogamic" with obvious flowers, or "cryptogamic" without apparent flowers, and along with such plants as Naiadaceae he groups the whole of the Vascular Cryptogams.

Apart from the fact that De Candolle's scheme is in many respects followed even yet in our modern floras—as, for instance, the subdivision of "Exogens" into Di- and Monochlamydeae and the further subdivision of the former into thalami-, calyci-, and corolli-florae—the system stood out as a protest against Linnaeus's artificial taxonomy, that found so much favour among the botanists of the early years of the nineteenth century, especially on the Continent. Unsatisfactory as it was from our point of view, it yet formed the starting-point of the improved scheme put forward by Bentham and Hooker, published more than fifty years after. (I will refer to his share in the great *Prodromus* later on.)

Beyond the general morphological principles laid down by A. P. de Candolle in his *Théorie Élémentaire*, nothing of any importance was added to our knowledge of the external morphology of plants during the first twenty years of the century. Considerable attention was, however, directed to internal structure, and many of the errors and misconceptions of Grew and Malpighi were corrected or removed. This was due no doubt, in the first instance, to the gradual improvements that were being introduced into the mechanism of the microscope, and also in large measure to the very considerable increase in the numbers of those who were attracted to anatomical studies and to the keen rivalry existing among them. This rivalry was accentuated by the offer of the Royal Society of Göttingen, in 1804, of a prize for the best essay on certain disputed points in plant histology. Three competitors entered for the prize, Rudolphi, Link, and Treviranus. It would serve no useful purpose to recount the views expressed in these three essays, or to describe the results arrived at by Bernhardi, who wrote a paper on plant vessels about the same period. How crude and even erroneous the conceptions of these days were you may judge by the fact that Bernhardi mistook pits for thickenings (evidently he had not learnt how to focus his microscope). Rudolphi expressed his belief in spontaneous generation, denied that fungi and lichens were plants, and thought that stomata were surrounded by sphincter muscles, while Treviranus held that intercellular spaces were filled with sap. Still there were a few points in which knowledge of anatomical structure was advanced. Thus Bernhardi recognised that spiral vessels had walls of their own and were not merely spirally wound fibres lying in a matrix; Link asserted that cells were closed vesicles and did not communicate with each other as Rudolphi thought. He also rejected Rudolphi's views on stomata and said that the apertures were surrounded by a cell or cells. Treviranus pointed out that pitted vessels were formed by the fusion of long cells placed end to end, and that all secondary thickenings were laid down on the inside of the thin walls of the elongated cells. He had also to his credit the discovery of stomata on the capsules of mosses.

Another anatomist, who was also a contemporary of those I have just mentioned, was Mirbel, who, among other publications, brought out a treatise in 1808 called *Exposition et défense de ma théorie de l'organisation végétale*. The theory he sets out to explain and defend was really adopted from Caspar Wolff, whose name I mentioned to you as that of the author of the *Theoria generationis*, published in 1759. Mirbel's version of Wolff's theory is that the precedent of all tissue elements is a homogeneous matrix in which cells appear as minute hollow cavities, and that tracheae are spirally wound laminæ inserted in the matrix. These elements are perforated by innumerable, but invisible, pores for the passage of sap from place to place. Mirbel identifies laticiferous tubes, but he confounds them with resin ducts, which you know to be intercellular spaces, and both of them with the true vessels of the xylem. In his pages we meet with the old misconception about wood being derived from the rind, while between them lies a nutrient sap in which new cells and tubes are to be found in various stages of formation.

Four years later, in 1812, a much more important work—*Beiträge zur Anatomie der Pflanzen*—was published by Moldenhawer—important not only on account of the many new histological facts that are recorded in it, but also because it formed the basis of the investigations of one of the greatest of the nineteenth-century anatomists, Hugo von Mohl, with whose work we shall have to deal in due course. To begin with, Moldenhawer introduced a new method in his investigations, viz. that of maceration of tissues in water. By this means he was able to isolate cells and fibres and examine them separately under the microscope. He then saw that the views of Wolff, Mirbel, and others, who held that these units appeared like bubbles in a uniform matrix, were quite untenable. The cavities of the cells were seen to be separated from each other by two walls, not one. He was able to demonstrate that fibrous elements were conjoined with vessels in long strands, separated by parenchyma, and thus introduced the term "fibrovascular bundle," so familiar to you. This discovery had far-reaching consequences, for it enabled Moldenhawer to present a new conception of the architecture of the plant organs. The stem of a Dicotyledon was no longer to be regarded as composed of rind, wood, and pith, but of vascular bundles, at first distinct from each other and gradually fusing as age increased. Increase in thickness is thus centred in the vascular bundle, and the ancient idea of a transformation of rind into wood is finally got rid of. Cells, he says, have no holes in their walls, and for the first time a stoma is described correctly.

It was almost inevitable that he should fall into some

errors. For one thing, he failed to differentiate between vessels proper, latex tubes, and resin passages, and, what was even worse, he thought the secondary thickening in vessels was laid down on the outsides of the walls, although Treviranus had already shown that that was not the case. With all its defects Moldenhawer's paper was a great advance on the productions of Mirbel, Link, Rudolphi, and others whose names I have mentioned to you as among his contemporaries.

Progress in Plant Physiology

Although substantial progress had thus been made in elucidating the intimate structure of plants during the early years of the nineteenth century, it is again in physiology that the greatest advances are seen, and these advances were in large measure due to the work of two men—De Saussure in nutrition and Knight in sensitivity.

In De Saussure's memoir, *Recherches chimiques sur la végétation*, published in 1804, we meet for the first time with quantitative as well as qualitative methods of investigation, and rigid experimentation on what might almost be regarded as modern laboratory lines. I think I cannot do better than give you some of De Saussure's conclusions in his own words. "When green plants," he says, "are exposed in atmospheric air to the successive action of day and night they inspire and expire alternately oxygen mixed with carbonic acid gas. The oxygen which green plants inspire is not directly assimilated by them; it is changed on inspiration into carbonic acid gas. They decompose this gas in the act of expiration, and it is only by this decomposition (which is only partial) that they are able to assimilate the oxygen which is present in the atmosphere. Roots, duramen, alburnum, petals, and all parts which in general are not green do not exhibit these successive inspirations and expirations ; they do not assimilate the oxygen either directly or indirectly ; they change it into carbonic acid gas, which is found in small quantities, stored or dissolved in the succulent parts as it would be in pure water, otherwise they do not alter it." As to the source of carbon, De Saussure claims that, whether cultivated in water or in air, plants obtain all their carbon from the carbon dioxide normally present in the atmosphere, and records a series of experiments to show that pea plants in an atmosphere containing about 8 per cent of carbon dioxide gained considerably in weight when exposed to direct sunlight, but that when the light was of feeble intensity the slightest addition of carbon dioxide to the air was detrimental. De Saussure was quite familiar with the importance of the green pigment, but he fell into the error of thinking that other pigments could take its place. Thus he experimented with plants that had red or purple foliage, and observed that such plants also gave off oxygen, and jumped to the conclusion that the green pigment was not essential to the decomposition, not noticing that it was actually present though masked by coloured cell sap.

De Saussure drew attention to the fact that in daylight a plant usually reassimilates all the carbon dioxide it has formed in the process of respiration, and hence that that phenomenon is not readily demonstrable while carbon assimilation is going on. He expressed astonishment that plants make no use of carbon monoxide or any other gaseous compound of carbon. A further point of great importance on which he lays stress is that neither carbon dioxide nor water are decomposed apart from each other, a discovery the significance of which was not appreciated for very many years. The extent of the advance that had been made in this relation you will better apprehend if you recall the fact that Ingen-Housz looked upon water merely as a vehicle for the transport of nutriment from the roots to the leaves. It was left for De Saussure to show this additional and important part played by water in the plant's nutritive economy.

De Saussure also showed that growth was impossible without respiration, and that the more active growth was, the more vigorous also was respiration. He found that no growth took place and that nutrition was abnormal in absence of minerals, and hence that these were not accidental impurities but essential constituents of the raw materials of the plant's food. His work contains a large number of analyses of plants and of soils, and he presents us with a long series of what might be termed balance-sheets of the income and outgo in various plants, and thus he arrived at a tolerably clear conception of what was necessary for adequate plant nutrition.

One very important conclusion he reached was that the nitrogen in the plant's composition did not come from the air but from the soil. He went astray somewhat in determining the precise source of the nitrogen, for he conceived it as being derived from animal and plant waste or from ammoniacal compounds formed during the decomposition of such waste. This was indeed the first beginnings of the so-called "humus theory" that retained its hold on the minds of plant physiologists for many years, until it was finally disproved by Boussingault about the middle of the century.

There is one other point which De Saussure emphasised, viz. that roots absorbed far more water than was required by the plant for nutritive purposes, and that the salt solution taken up was extremely dilute; but his ignorance of the law of osmosis prevented him from offering any appropriate explanation of this phenomenon. He noticed, however, that roots absorbed substances in solution whether these substances were of service in nutrition or not, thus contradicting the older view that roots had the power of selecting from the soil just what they required.

To sum up, it may be said that the chief merit of De Saussure's work lies in the rigidly scientific methods he employs in all his researches, the logical manner in which he marshals his facts, and the careful deductions he draws from them. Doubtless his mental attitude owed much to his intimate association with his father, the distinguished physicist and Alpine climber, and perhaps even more to his thorough study of the writings of Lavoisier. By many De Saussure is spoken of as the last of the little band of investigators in plant physiology who flourished at the end of the eighteenth and in the very early years of the nineteenth centuries. I think it would be more correct to regard him as the forerunner of the new school of chemical biologists represented more than a generation afterwards by Boussingault and Liebig. In any case, in whichever category we elect to place its author, there can be no doubt that in De Saussure's treatise we have a classic well worthy to rank alongside the *Vegetable Staticks* of Hales and the *Experiments on Vegetables* of Ingen-Housz.

Another outstanding feature in the department of physiology in the early years of the nineteenth century is the work of Thomas Andrew Knight. His earlier investigations dealt with the circulation of sap, but there he was obsessed with the idea that the sap of plants corresponded to the blood, and that therefore there must be a circulation in the plant comparable with that seen in the animal. Notwithstanding Ingen-Housz's work on the subject, Knight still held to the notion that the sap, after having reached the leaves *via* the vessels of the wood, and after having been exposed there to light and air, acquired, " by means I shall not attempt to decide, the power to generate the various inflammable substances that are formed in the plant. It appears to be then brought back again through the vessels of the leaf stalk to the bark, and by that to be conveyed to every part of the tree to add new matter, and to compose its various organs for the succeeding season." Knight even went the length of resuscitating the ancient idea that there were valves in the bast vessels for regulating the sap flow. He traced the pathway of ascent of sap by Hales's method of ringing, and by immersion of the cut ends of branches in an infusion of grape skins. He thought that "perspiration" (transpiration) was effected directly by the under surface of the leaf while the upper surface absorbed water. Most of the water came from a reservoir in the pith, so that his knowledge of root absorption cannot be regarded as very accurate or profound. The old notion of peristalsis in the wood vessels as a cause of the ascent of sap reappears in Knight's papers, for he suggests that the movement may be due to rhythmic expansion and contraction of the "silver grain" under the influence of heat.

In the discussion of these and similar problems in nutrition Knight is far behind Hales, who lived almost a century before his day. But in one section of physiology, sensitivity, he made experiments that are still quoted, and drew deductions from them that are incorporated in all our present-day accounts of the phenomena of stimulus and response. At the same time you must not imagine that Knight had any conception of plant sensitivity as we understand the term. Geotropic and heliotropic curvatures were to him simple bending movements that had to be explained mechanically; protoplasm with all its mysterious capabilities for appreciating and responding to stimuli was to Knight a sealed book.

In a paper read to the Royal Society in 1806 he opens the question by stating that while studying seed germination he had observed that no matter how he placed the seed "its radicle invariably makes an effort to descend towards the centre of the earth, while the elongated germen (plumule) takes a precisely opposite direction." He then hints that "some naturalists have supposed these opposite effects to be produced by gravitation." Knight, however, was not content to let the matter rest there; he set himself the task of finding out whether gravitation had or had not anything to do with the bendings. "As gravitation could produce these effects," he says, "only while the seed remained at rest and in the same position relative to the attraction of the earth, I imagined that its operation would become suspended by constant and rapid change of the position of the germinating seed, and that it might be counteracted by the agency of centrifugal force." He then attempted to prove his theory experimentally by placing germinating seeds on the circumference of a wheel which was made to revolve with great rapidity in a vertical plane. "The radicles of these seeds," he continues, "were made to point in every direction, some towards the centre of the wheel and others in the opposite direction; others as tangents to its curve, some pointing backwards and others forwards, relative to its motion; and others pointing in opposite directions in lines parallel with the axis of the wheel. I had soon the pleasure to see that the radicles in whatever direction they were protruded from the position of the seed, turned their points outwards from the circumference of the wheel, and in their subsequent growth receded nearly at right angles from its axis. The germens on the contrary took the opposite direction, and in a few days their points all met in the centre of the wheel."

Knight then altered the position of his wheel so as to make it revolve in a horizontal plane at different rates, trying in this way to find what would be the result of combining gravity with centrifugal force. "The difference I had anticipated between the effects of rapid vertical and horizontal motion soon became sufficiently obvious. The radicles pointed downwards about 10° below, and the germens as many degrees above, the horizontal line of the wheel's motion; centrifugal force having made both to deviate 80° from the perpendicular direction each would have taken, had it vegetated at rest. Gradually diminishing the rapidity of the motion of the horizontal wheel, the radicles descended more perpendicularly, and the germens grew more upright ; and when it did not perform more than eighty revolutions in a minute the radicle pointed about 45° below, and the germen as much above, the horizontal line, the one always receding from, and the other approaching to the axis of the wheel."

It is curious to note how Knight fights shy of attributing any power of perception—"sensation," he called it—to the radicle and plumule ; he feels himself obliged to hunt for a mechanical explanation of the phenomena, and here it is—a rather lame one, it must be admitted. "The new matter which is added [to the root] unquestionably descends in a fluid state from the cotyledons. On this fluid, and on the vegetable fibres and vessels while soft and flexible, and whilst the matter which composes them is changing from a fluid to a solid state, gravitation, I conceive, would operate sufficiently to give an inclination downwards to the point of the radicle. The germen elongates by a general extension of its parts previously organised, and if the motion consequent to distribution of the true sap be influenced by gravitation, it follows that when the germen deviates from a perpendicular direction the sap must accumulate on its under side ; the fibres and vessels on the under side of the germen invariably elongate much more rapidly than those on its upper side ; and thence it follows that the point of the germen must always turn upwards." You will see that Knight is forced to admit that the mechanical explanation, though it might account for the downward growth of roots, will not account for the upward growth of the plumule without dragging in an elongation of the vessels and fibres of the under side. It is strange that he did not see that the same reasoning might be applied to a root lying horizontally, and yet in spite of the sinking of the sap to its under side the root invariably bends downwards. It does not appear to have occurred to him to ask why the vessels and fibres of the under side of the shoot should elongate and not the vessels and fibres of the root also.

During his studies Knight observed that roots were attracted by moisture, and, by a series of experiments devised for the purpose, he showed that water also could counterbalance the action of gravity or, as we would say, hydrotropism could overcome geotropism. Once more Knight misses the idea of the sensitivity of the living organism to an external stimulus. These experiments were detailed in a paper published in 1810.

A third series of experiments were carried out during the next two years on the apheliotropic movement of the tendrils of the vine and Virginian creeper and the aerial roots of ivy. The explanation he offered, however, still exhibits his persistent adherence to a mechanical explanation of these responses to stimulus. "The external pressure," he writes, "of any body upon one side of a tendril will probably drive the fluid organisable matter from one side of the tendril, which will consequently contract, to the other side, which will expand, and the tendril will thence be compelled to bend round a slender bar of wood." The curvature, you see, is, according to Knight's idea, a passive result of the transference of fluid from the pressed to the free side, not an active curving in response to the pressure.

Knight was also struck by the way in which leaves had a tendency to arrange themselves so that the incident light always fell on the upper surface, or, as we would put it, that leaves were diaphototropic. "I will request your attention," he writes, "to the power of moving in the Vine leaf, on which I have made many experiments. It is well known that this organ always places itself so that the light falls upon its upper surface, and that if moved from that position it will immediately endeavour to regain it; but the extent of the efforts it will make, I have not anywhere seen noticed. I have very frequently placed the leaf of a vine in such a position, that the sun has shone strongly on its under surface; and I have afterwards put obstacles in its way on which ever side it attempted to escape. In this position the leaf has tried almost every method possible to turn its proper surface to the light." Do you notice how close Knight gets to the true reason for these movements? Observe his phraseology—"the leaf tried almost every method possible." The words suggest at once a conscious effort on the part of the leaf to do something it was prevented from doing by some external agent. Why did not Knight attribute to the leaf the power of perception or "feeling" so clearly suggested by the very words he uses to express the leaf's activities? But no, he will have none of it! "I am wholly unable to trace the existence of anything like sensation or intellect in plants," he says, and yet the thing was staring him in the face! "As the whole effort here produced appears to arise merely from the light falling on the under surface of the leaf, I cannot conceive how the contortions of its stalk, in every direction, can be accounted for without admitting not only that the leaf possesses an intrinsic power of moving, but that it also possesses some vehicle of irritation." That is precisely what the plant does possess, although Knight's eyes were blinded to it. Yet he saw much that was hidden even from his successors in the same field of enquiry, and what he saw and recorded entitles him to our admiration and respect

LECTURE V

ROBERT BROWN

THE outstanding figure in the history of botany during the twenty years succeeding Knight was that of Robert Brown, the son of a Scottish clergyman of Montrose. After a period devoted to the study of medicine and, in his leisure moments, to an exploration of the Scottish flora, he joined the army, but ultimately left it to accompany a surveying expedition to Australia. During the four years Brown spent in the antipodes he gathered together an enormous amount of material representative of the then almost unknown flora of Australia, and during his voyage and for five years after his return home he worked without cessation at the description and arrangement of his collections. He was fortunate in having gained the warm friendship and invaluable interest of Sir Joseph Banks, the famous explorer, and had also the inestimable advantage of having free access to all the collections, including that of Linnaeus, in the custody of the Linnean Society, whose librarian he became soon after his return to England.

Among the first fruits of his labours there appeared in 1810 a *Prodromus* or preliminary account of the flora of New Holland and Tasmania. The Linnean system, then in the heyday of its favour among botanists, had obviously no attraction for him, for the arrangement adopted in the *Prodromus* is pronouncedly De Candollean, although he remodelled several of De Candolle's families and added new ones to accommodate the numerous new types he had discovered in his travels. He also wrote a general account of the botany of the Australian continent and compared the flora with that of South Africa, South America, and other regions of the Southern Hemisphere, and thus laid the foundations of botanical geography, a subject destined to develop so markedly in the hands of Sir Joseph Hooker a generation afterwards.

About the same time Brown began the series of publications on special genera and groups for which he is most famous, and the first of these was a monograph on the characteristically Australian family of the Proteaceae. It has often been pointed out by his biographers and critics that Brown had a curious habit of hiding away some of his most important morphological and anatomical discoveries in the body of an otherwise purely descriptive or taxonomic paper. This monograph on the Proteaceae is a case in point. After a full account of the genera and species of the order, with remarks on their affinities and geographical distribution, he quite incidentally enlarges on the structure of the seed in general, and elucidates in a sentence the morphological nature of the seed reserves. "The albumen of the seed," he says, "is merely that condensed portion of the liquor amnios which remains unabsorbed by the embryo; and as this fluid is in the early stage never wanting, all seeds may in one sense be said to have albumen; but while in some tribes this unabsorbed part in the ripe seed many times exceeds the size of the embryo, so there are others in which not a vestige of it remains." It seems to me that this brief statement of the relations of the endosperm to the embryo gives a clearer exposition of the events taking place in the process of seed development and maturation than is to be found in very many of the botanical textbooks in use at the present moment. In this paper also Brown goes minutely into the structure of the pollen grains in the Proteaceae and their adaptations to the morphological features of the stigma.

The *Prodromus* also contains valuable morphological

observations on the flowers of the Polygalaceae, and the limits of this order are defined anew as based on these observations. Similarly, in the general account of Australian botany I have just referred to, he worked out the true morphology of the curious and deceptive inflorescence of the Euphorbiaceae and the almost equally puzzling inflorescence of the grasses.

During the years that followed, the newly made collections of travellers, more especially on the African continent, were sent to Brown for report, and the results of his labours on them were embodied in a succession of memoirs on the same lines as those he had followed in dealing with the Proteaceae. Many well-known groups were analysed and put into proper order, more especially the Cruciferae, Capparidaceae, Resedaceae, Leguminosae, and Compositae. In all these researches and many others of a like nature Brown did immense service to the cause of the "Natural System," in the first place by invariably discussing the orders from that point of view only and, without actually attacking the Linnean system, dealing it a crushing blow by simply ignoring it altogether, incidentally bringing out in his own treatment of the orders the fundamental correctness of the natural system.

Although invited to occupy the chair of botany both at Edinburgh and Glasgow Universities, Brown preferred to remain in London and, in the seclusion of his study, to continue to work unobtrusively at the collections put in his charge until the day of his death in 1858.

One of the best known papers that came from Brown's pen during this period was a monograph on the genus Kingia, belonging to the Liliaceae. As an appendix to it he printed an account of his investigations into the structure of the ovule in flowering plants in general, and also a discussion of the reproductive organs of Cycads and Conifers. In the first of these contributions Brown made some remarkable statements. He described the ovule as consisting of a "nucleus" (nucellus) covered by two integuments, save at the apex, where a micropyle was left to facilitate impregnation. The vascular supply in the raphe, the formation of the endosperm and the relation of the embryo to that and other morphological structures, are all described, save for small differences in terminology, exactly as you would be expected to describe them to-day in an examination paper. His investigation of the female cones of the Cycadaceae and Coniferae led him to believe that in some cases the ovules might not be enclosed in carpels and that the pollen might reach the micropyle directly without being first received by the stigma. These considerations induced him to regard these orders as naked-seeded, or gymnospermatous. He also recognised polyembryony in the Gymnosperms, although he did not publish any observations on the subject until long afterwards. He found that the nucellus of the ovule must be distinguished from what he termed the "amnios" or endosperm, and within the amnios he identified three or more clearer regions, each after fertilisation containing a branched thread or threads, at the free ends of which the embryos were developed. He termed these clearer regions "corpuscula," adopting the name from the French botanist Du Petit Thouars, who had also observed them in Cycas.

You will recollect how, at the end of the eighteenth century, the subsequent history of the pollen grain after it had reached the stigma was left undecided, and how it was very generally believed that the embryo was present in the ovary previous to pollination and merely stimulated to renewed activity in some mysterious way by an oil or other excretion formed on the stigma. Hazy ideas of this kind were quite repugnant to a man of Brown's temperament, who, in all his work, insisted on accurate observation as opposed to vague theorising. An opportunity presented itself to him in 1831 to solve the riddle while examining the flowers of the Orchidaceae, which, as you are aware, have pollinia or massed pollen grains. After working out the morphology of the flower and showing, with his usual acumen, how it is related to the typical trimerous flower of the Monocotyledons, he found that insects in their passage from flower to flower carried the whole pollinium with them and that several stigmas might be pollinated in succession by the same pollinium. He next noticed the formation of the pollen tubes from the pollen grains on the stigma and traced them right down to the placenta, but there he lost them. The extrusion of the pollen tube from the grain had already been noticed by Amici in 1823 and by Brongniart in 1826, but the significance of the phenomenon had not been appreciated. Brown found tubes similar in appearance entering the micropyles, but he could not convince himself that these were continuations of the pollen tubes, though he conjectured that they were at least derived from them. He accounts for the protrusion of the pollen tube from the grain by assuming that it received a stimulus from the stigma, and thinks that the tube during its passage to the ovary is nourished by something derived from the tissue of the style. This nutritive relation between pollen tube and style was worked out sixty years later by Reynolds Green.

In the course of a corresponding enquiry into the structure of the flower of Asclepias he carried the matter a step farther by tracing the connection between the pollen tubes and the tubes he had seen entering the micropyles of the orchid ovules, and also recognised granules inside these tubes, although he thought they were concerned with nutrition and not with fertilisation. It is in his paper on the Orchidaceae that he made the great discovery that was destined in after years to lead to the establishment of a new section of plant histology, viz. Cytology. This observation of Brown's is of so much interest that I will quote you the sentence in which he announces his discovery: " In each cell of the epidermis of a great part of this family (Orchidaceae), especially of those with membranaceous leaves, a single circular areola, generally somewhat more opaque than the membrane of the cell, is observable. This areola, which is more or less distinctly granular, is slightly convex, and although it seems to be on the surface is in reality covered by the outer lamina of the cell. There is no regularity as to its place in the cell; it is not infrequently, however, central or nearly so. . . . This areola, or nucleus of the cell, as perhaps it might be termed, is not confined to the epidermis, being also found in the parenchyma or internal cells of the tissue. In the compressed cells of the epidermis the nucleus is in a corresponding degree flattened, but in the internal tissue it is often nearly spherical, more or less firmly adhering to one of the walls and projecting into the cavity of the cell. . . This nucleus of the cell is not confined to Orchideae, but is equally manifest in many other Monocotyledonous families; and I have even found it, hitherto however in very few cases, in the epidermis of Dicotyledonous plants." In these simple words Brown announced the discovery of the cell nucleus on which a whole library of memoirs was to be written in the years to follow, both on its minute structure and on the complex changes it undergoes during the process of cell division.

There is another but in itself relatively unimportant observation made by Brown and now always known by his name, viz. "Brownian movement." This term, as you may know, is applied to the oscillation of minute particles in a fluid as seen under the microscope. These movements are purely physical, as indeed Brown himself ultimately concluded, seeing that they take place among inorganic particles of sufficiently small size when suspended in water. Various explanations have been offered to account for this vibration, the one commonly accepted being that it is due to the perpetual bombardment of these particles by the molecules of the medium, but for this and other theories I must refer you to works on physical chemistry.

You will now, I hope, have begun to appreciate how much botany owes to Robert Brown's investigations. In almost every department of the science he left his mark, and if he did not himself promulgate a complete system of classification, he did what was of even greater service in this transitional period, he prepared and published many monographs on individual orders and compared them with their nearest allies, thus demonstrating within narrower cycles of affinity the broader principles that he regarded as those that should guide future progress towards the goal that all true taxonomists saw ever ahead, viz. the formation of a genuine natural classification founded on racial affinity and blood relationship.

The great American taxonomist, Asa Gray, summed up Brown's characteristics in the following striking words: "Brown delighted to rise from a special case to a high and wide generalisation, and was apt to draw most important and always irresistible conclusions from small selected data, or particular points of structure, which to ordinary apprehension would appear wholly inadequate to the purpose. He had unequalled skill in finding decisive instances. So all his discoveries, so simply and quietly announced, and all his notes and observations, sedulously reduced to the briefest expressions, are fertile far beyond the reader's expectation. Cautious to excess, never suggesting a theory until he had thoroughly weighed all the available objections to it, and never propounding a view which he did not know how to prove, perhaps no naturalist ever taught so much in writing so little, or made so few statements that had to be recalled or even recast."

On Robert Brown, also, the distinguished naturalist Von Humboldt conferred the title so well merited as to receive universal confirmation by all his fellow botanists:

"Facile botanicorum princeps; Britanniae gloria et ornamentum."

During the first decades of the nineteenth century quite a number of attempts were made to classify plants on what were believed by their authors to be sound natural principles, but none of these were successful in displacing the De Jussieu-De Candolle system. Endlicher, about 1836, put forward a scheme based on totally erroneous ideas as to the modes of growth of different types of plant life, a scheme which very soon died a natural death. Similarly Brongniart, whose work in another department of the science I shall have to speak of by and by, divided plants principally into Cryptogams and Phanerogams, including under the latter Gymnosperms, Monocotyledons, and Dicotyledons. He prefixed a group, Agames, in which he placed Algae, Fungi, and Lichens, whose sexual organs were at that time, with very few exceptions, practically unknown. On the whole the classification, which appeared in 1828, was an advance on those then before the botanical public, and possessed the merit of placing the Gymnosperms in a class apart from the Dicotyledons, with which some botanists long after Brongniart's day persisted in classing them.

Lindley was another systematist who flourished in the early years of the nineteenth century. He was largely instrumental in preserving the Royal Gardens at Kew as a national institution when Parliament proposed to abolish them altogether. His services to horticulture were many and important; indeed, in the words of the President of the Royal Society when presenting Lindley with the Royal Medal, "he raised horticulture from the condition of an empirical art to that of a developed science." His most important work, *The Vegetable Kingdom*, was published in 1846, and in it he formulated the last of several schemes suggested by him. How far it was likely to succeed may be judged by the fact that he based it on a principle which he enunciates in the following words: "Physiological characters are of greater importance in regulating the natural classification of plants than structural," thus showing himself diametrically opposed to De Candolle, who asserted that a true classification must be founded on morphological features only. He divided plants into Asexual or Flowerless and Sexual or Flowering. The Asexual include Thallogens and Acrogens, while the Sexual embrace five classes: Rhizogens (parasites), Endogens, Dictyogens, Gymnogens, and Exogens. The Rhizogens had nothing in common beyond their parasitism—a physiological characteristic, while the Dictyogens were merely a few orders of Monocotyledons, or Endogens, with reticulate veined leaves. The Exogens or Dicotyledons are subdivided into diclinous and hermaphrodite, and the latter again into hypogynous, perigynous, and epigynous orders. I do not think it would serve any useful purpose were I to explain to you in further detail any of the systems Lindley put forward from time to time, still less to expose the many errors and inconsistencies found in them. Without actually going so far as to say, with Sachs, that Lindley's final effort at subdividing plants as given in the *Vegetable Kingdom* was "one of the most unfortunate classifications ever attempted," still the general impression one gets from a study of his publications on taxonomy is that his mind was in constant state of vacillation on the subject, laying down guiding rules which he does not follow or follows only half-heartedly, and abandoning one scheme only to replace it with another and more erroneous one.

You will remember that A. P. de Candolle had been the leading opponent of Linnaeus's artificial system and had put forward a classification of his own based on or adapted from those of Ray and De Jussieu. His son, Alphonse de Candolle, followed up his father's work and is mainly responsible for the great *Prodromus Systematis naturalis regni vegetabilis.* This immense work, which made its appearance gradually over a number of years, followed in the main the original lines laid down by the elder botanist. Cellular and Vascular Cryptogams are, however, united into a division equal in value to Phanerogams, and these latter are divided into Exogens (Thalamiflorae, Calyciflorae, Corolliflorae, and Monochlamydeae, in which last group are included Cycadaceae and Coniferae) and Endogens.

At this point I should like to refer, very briefly, to a series of morphological investigations which had nothing to do with taxonomic problems. The first of these was a research by K. F. Schimper, published in 1830, in which the author evolved a theory of the arrangement of leaves on the plant axis. The theory appealed very strongly to botanists with a mathematical bent of mind, for the conclusions that Schimper arrived at were expressed in formulae which appeared to be capable of being manipulated in accordance with certain recognised mathematical principles.

In studying the order of succession of leaves on a stem Schimper noticed that, in the simplest case, starting from leaf A, leaf B arose at the opposite side of the axis, but at a higher level, and that leaf C originated immediately above leaf A, leaf D above leaf B, and so on. There was thus a lateral divergence of 180 degrees between the points of origin of any two successive leaves, and if a line were drawn on the axis uniting the bases of these it described a spiral round the stem. The divergence in this, the simplest case, might be expressed by the fraction $\frac{1}{2}$. The numerator indicated the number of times the spiral line passed round the stem in its passage from the initial leaf to that immediately above it—in this case once—and the denominator the number of leaves passed in that progress—in this case two. The next instance examined was where there were three leaves arising from the axis on one turn of the spiral, with a divergence, therefore, of 120 degrees, a condition represented by the fraction $\frac{1}{3}$. The next condition was when two spiral turns had to be made, starting from the initial leaf and touching each successively higher leaf, before the leaf immediately above the initial leaf in the same vertical plane was reached, while five leaves had to be passed in the two circuits of the stem. This was expressed by the fraction $\frac{2}{5}$. It was then seen that by adding the numerators and denominators of any two successive fractions the next higher fraction in the series could be arrived at, thus $\frac{1}{2}+\frac{1}{3}=\frac{2}{5}$: $\frac{1}{3}+\frac{2}{5}=\frac{3}{8}$: $\frac{2}{5}+\frac{3}{8}=\frac{5}{13}$, and so on. A vast amount of ingenuity and skill was expended, both by Schimper and later by Alexander Braun, in working out this idea, and it was not long before the whole subject, with its cacophonous terminology of orthostichies, parastichies, and what not, became the new fad of the mathematically minded botanist. The relationship between the point of origin of the leaf, its form, and, most important of all, its functions passed into the background, and all efforts seemed to be focussed on the attempt to establish this pseudo-mathematical idea as one of the fundamental laws of nature. You can scarcely open one of the older textbooks of botany without meeting with pages of calculations and drawings of transparent stems, idealised pine cones, and machine-turned figures of capitula, etc., with various lines drawn upon them, illustrating the principles on which this new subject of phyllotaxis was supposed to be based. I have no desire to inflict on you the miseries I and my fellow-students endured in our endeavours to master what was regarded as one of the articles of a botanist's faith in the later years of last century.

I ought at this point to sketch for you the very important advances made in anatomy and histology during the period with which we are at present concerned, but I prefer to delay consideration of that aspect of the subject till the next lecture, so that I may present the story to you in a more connected form. Meanwhile let us enquire what the physiologists had been doing and how far they had made progress since Ingen-Housz and De Saussure had opened up new vistas in the subject, and set out new problems for solution. You will most readily gain an insight into the condition of plant physiology at or about the year 1830 by studying De Candolle's account as it appeared in his *Physiologie végétale*, published in 1832, and which I will quote to you in full. Here is his view of plant nutrition:

"The spongioles of the root" (in reality the growing apices) "being actively contractile and aided by the capillarity and hygroscopic qualities of their tissue, suck in the water that surrounds them, together with the saline, organic, or gaseous substances with which it is laden. By the operation of an activity which is manifested principally in the contractility of the cells and perhaps also of the vessels, and is maintained by the hygroscopic character and capillarity of the tissue of the plant and also by the interspaces produced by expiration of the air and by other causes, the water sucked in by the roots is conducted through the wood, and especially in the intercellular spaces, to the leaf-like parts, being attracted in a vertical direction by the leaves and in a lateral direction by the cellular envelope at every period of the year, but chiefly in the spring; a considerable part of it is exhaled all day long through the stomata into the outer air in the form of pure water, leaving in the organs in which the evaporation takes place all the saline, and especially all the mineral, particles which it contained. The crude sap which reaches the leaf-like parts of the plant there encounters the sunlight, and by it the carbonic acid gas held in solution by the sap, whether derived from the water sucked in by the roots or from the atmospheric air, or being part of that which the oxygen of the air produced with the surplus carbon of the plant, is decomposed in the day time; the carbon is fixed in the plant and the oxygen discharged as gas into the air. The immediate result of this operation appears to be the formation of a substance which in its simplest and most ordinary state is a kind of gum consisting of one atom of water and one of carbon, and which may be changed with very little alteration into starch, sugar, and lignine, the composition of which is almost the same. The nutrient sap thus produced descends during the night from the leaves to the roots, by way of the rind and the alburnum in Exogens, by way of the wood in Endogens. On its way it falls in with glands or glandular cells, especially in the rind and near the place where it was first formed; these fill themselves with the sap and generate special substances in their interior, most of which are of no use in the nutrition of the plant, but are destined either to be discharged into the outer air or to be conducted to other parts of the tissue. The sap deposits in its course the food material, which becoming more or less mixed up with the ascending crude sap in the wood, or sucked in with the water which the parenchyma of the rind draws to itself through the medullary rays, is absorbed by the cells, and chiefly by the roundish or only slightly elongated cells, and there further elaborated. This storing up of food material, which consists chiefly of gum, starch, sugar, perhaps also lignine, and sometimes fatty oil, takes place copiously in organs appointed for the purpose, from which this material is again removed to serve for the nourishment of other organs. The water, which rises from the roots to the leaf-like parts of the plant, reaches them in an almost pure state, if it passes quickly through the woody parts, the molecules of which are but slightly soluble. If, on the other hand, the water flows through parts in which there is much roundish cell tissue filled with food material, it moves more slowly and mixes with this material and dissolves it; when it is drawn away from these places by the vital activity of the growing parts, it reaches them not as pure water but charged with nutrient substances. The juices of plants appear to be conveyed chiefly through the intercellular passages. The vessels probably share in certain cases in these functions, but serve generally as air canals. The cells appear to be the really active organs in nutrition, since decomposition and assimilation of the juices takes place in them. Cyclosis is a phenomenon which appears to be closely connected only with the preparation of the milky juices, and to be caused by the actively contractile nature of the cell walls or of the tubes. Woody and other substances are deposited in every cell in different quantities according to their kinds and the accompanying circumstances, and clothe their walls; the unequal thickness of the layer so deposited appears, according to Hugo von Mohl, to have given rise to the supposition of perforated cells; that is, the parts of the cell wall that remain transparent appear under the microscope as pores. Every cell may be regarded as a body which prepares juices in its interior; but in vascular plants their activity stands in such a connection with a complex of organs, that a single cell does not represent the whole organism, as may be said of the cells of certain cellular plants, which are all like one another. There is no circulation in plants like the circulation in animals, but there is an alternating ascent and descent of the crude sap and of the formative sap which is often mixed with it. Both these phenomena depend perhaps on the contractile power in cells that are still young, and, if so, this power should be the true vital energy in plants" (Sachs).

Such is De Candolle's version of plant nutrition. Disappointing, is it not, after all the work of Ingen-Housz and De Saussure? Here are the same old absurdities and misconceptions with new ones added. The "spongioles" —really the swollen meristematic ends of roots with their protective root caps—are special inventions of De Candolle, which, like many other erroneous notions, died hard.

But you will recognise several old friends such as contractility of cells and vessels, intercellular spaces as sap conduits, soil as the source of some at least of the carbon dioxide, the ascent of " crude sap " by day and the descent of " elaborated sap " by night, the muddle between reserves and excreta, not to speak of the " vital energy " which was supposed in some mysterious way to be associated with the contractility of young cells. All these blunders and misconceptions, in a treatise on vegetable physiology written a generation after the days of De Saussure, make one marvel how such a pronouncement could have emanated from the pen of a botanist of the rank of De Candolle. It is indeed but a short step in advance of the phantasies of the sixteenth-century herbalists.

Another physiological investigator, a countryman and a contemporary of De Candolle, was Dutrochet. In his *Mémoires*, published in 1837, Dutrochet cleared up quite a number of special points. He was the first to emphasise the importance of osmosis in plant nutrition, a subject he had studied some years previously in relation to the escape of gonidia in Thallophyta. He distinguished " spring bleeding," which he attributed to root pressure resulting from excessive endosmosis, from the normal transpiration current, which he regarded as induced by the suctional power of transpiring leaves, the *vis a tergo* and the *vis a fronte* of the early textbooks. He also did good service in pointing out that respiration in plants is identical with that in animals, and thus was instrumental in destroying the old notions about diurnal and nocturnal respiration. On respiration, according to Dutrochet, all growth and movement were absolutely dependent, and he further showed that the evolution of heat, with or without an accompanying rise of temperature, was a necessary consequence of the chemical processes taking place in all active plant organs. The recognition of these chemical phenomena in plants as well as the synthesis of organic compounds which, just before this date, had been effected by Wöhler, sounded the death-knell of the " vital force " that had dominated the minds of physiologists for over half a century. After 1840 we hear no more of this *deus ex machina* that was dragged in to account for everything that could not be explained by a reference to the known laws of physics and chemistry.

Another name came very prominently to the front about 1840, viz. that of Liebig, who wrote a famous treatise entitled *Die organische Chemie in ihrer Anwendung auf Agrikultur und Physiologie.* Liebig was rather an agricultural chemist than a botanist, and was therefore primarily interested in the nature of the soil constituents and the use the plant made of them. One of the greatest services he rendered to botany was the refutation of the " humus theory," so tenaciously held by physiologists and agriculturists alike. This theory, as you doubtless remember, postulates that the chief source of the supply of carbon and other materials to the plant is the waste from the animal and vegetable kingdoms. Liebig exploded this view by showing that humus was not absorbed by roots and that, so far from reducing it in quantity, as would naturally be the case if plants lived upon it, the plant actually added to it. He further showed that even if it were absorbed, the amount present in the soil was quite insufficient to provide for the wants of vegetation. On the contrary, he insisted that the only source of carbon available to the plant was the carbon dioxide of the air. (Yet some years after Schleiden wrote " that the leaves in their natural growth absorb carbonic acid from the air was a pure invention " !) He also asserted that " carbon dioxide, ammonia, and water contain in their elements the requisites for the production of all the substances that are in animals and plants during their lifetime. Carbon dioxide, ammonia, and water are the ultimate products of the chemical process of their putrefaction and decay." This important generalisation was confirmed and extended by Boussingault ten years later, when he showed that plants were unable to make any use of the free nitrogen of the air, but derived all their supplies of that element from nitrates present in the soil, and that plants could be cultivated quite satisfactorily in an artificial soil containing nitrates but from which humus had been entirely excluded. To complete the story, it was shown, about the same time, by Wiegmann and Polstorff that the minerals present in plant ash were essential to the welfare of the plant and were all derived from the soil and from the soil only.

Cryptogamic Botany in the beginning of the Nineteenth Century

When we were discussing the evolution of the idea of a natural classification of plants you must have noticed how the lower forms of vegetable life were practically ignored, or bundled together, often along with some Phanerogams whose flowers could not be seen or at least could not be recognised as such, into a general rag-bag called Cryptogamia, Cellulares, Acotyledons, and so on. The attention of taxonomists was concentrated on flowering plants and, save for a few isolated observations made during the closing years of the eighteenth and early years of the nineteenth centuries, very little was known of either the structure or the life histories of the Thallophyta, Bryophyta, and Vascular Cryptogams. As recently as 1881 the course of instruction in botany in at least one British University consisted of fifty lectures, forty-five of which were devoted to the morphology, physiology (very little of it!), and taxonomy of flowering plants, while the remaining five dealt with the life history of a fern and of a moss and the merest sketch of the differences in nutrition between an Alga and a Fungus. In the first half of the eighteenth century, it is true, some observations were made on Fungi and their mode of development, but no comprehensive treatise dealing with the lower forms of plant life came into existence till well on in the nineteenth century, and any work actually accomplished before that date was concerned chiefly with the occurrence or non-occurrence of sexuality among such organisms.

After 1840 or thereabout sexuality in flowering plants was universally accepted, since Amici's work in 1846 completed the story, for the earlier chapters of which he and Robert Brown had been responsible. Schleiden made an attempt in 1837 to show that the embryo was formed from the apex of the pollen tube and that the ovule acted merely as a nidus or nutrient bed for its further development, but this absurd idea was soon disproved by Amici, who, in 1842 and 1846, demonstrated in the embryo-sac the presence of an ovum which was fertilised by a "fluid" extruded from the tips of the pollen tube. (Remember that the profound significance of the nucleus was not at that time appreciated.) In 1856 Radlkofer finally gave Schleiden's hypothesis the "coup de grâce."

In 1803 the Swiss theologian Vaucher noted conjugation in some of the lower green Algae, and suggested that the fusion might be regarded as a sexual act, while phenomena of a similar nature were also observed in the lower Fungi. Several writers also recorded the occurrence of sperms in Chara, Liverworts, and Mosses. Hedwig, who was the first to study this last group at all carefully, expressed the opinion that the antheridia and archegonia in these plants corresponded to the stamens and carpels of the Phanerogams. Similar organs were discovered by Suminski, in 1848, on the prothalli (then spoken of as cotyledons) of ferns, while some pioneer work was accomplished on Fucus by Thuret in 1845, and on Rhodophyceae by Naegeli in 1846. Looking at the subject broadly, however, it may be said that it was not until well after the middle of the century that Cryptogams received that attention to which they were entitled. For the present therefore we may leave the matter on one side and turn to the consideration of two subjects of fundamental importance in the history of Biology, viz. the discovery of protoplasm and the foundation of the cell theory.

Protoplasm and the Cell Theory

Let me recall to your memory the condition in which we left the conception of plant movement. Knight, in 1812, after examining with wonderful skill the mysterious performances of roots and stems in relation to gravity, and the equally puzzling contortions of vine leaves in their efforts to regain their original orientation with reference to incident light, finished up with the confession that he could not explain the movements "without admitting not only that the plant possesses an intrinsic power of moving but that it also possesses some vehicle of irritation."

In the years following the publication of Knight's classic researches numerous additional instances of movement were recorded, and also many new observations on such movements as had already been recognised as taking place under certain conditions. That roots when placed in a horizontal position did not bend downwards simply on account of their own weight became evident after it had been demonstrated that they could push down the scale-pan of a balance against the resistance of a weight in the other pan, and could even force their apices into mercury. Dutrochet and Von Mohl went the length of ascribing to tendrils, twining stems, and mobile leaves a certain degree of sensitiveness, and regarded the contact with the foreign body as the stimulus. Dutrochet also found that some stems, such as those of parasites, were negatively heliotropic while others were positively geotropic. In short, evidence was rapidly accumulating tending to show that these movements were not purely physical or mechanical, but were of the nature of responses to external stimuli, and that the precise character of the response depended on the plant organ and not on the stimulus. What then was it that was sensitive or, in Knight's phraseology, what was "the vehicle of irritation"?

Hooke, you will remember, was the first, in 1665, to employ the term "cell," but he applied it to the minute cavities he discovered in vegetable tissues, while the walls were regarded by him and others after him as composed of an "interstitial substance." As the microscope improved, the attention of the early histologists became more and more directed to discovering the nature of these "interstitia," until Moldenhawer, by his maceration method, finally demonstrated that each cell and fibre had a wall of its own. Then arose a long controversy on the nature, origin, and mode of thickening of the wall, a subject we shall have to discuss presently. The cell wall now came to be looked upon as the "cell" instead of only something separating two or more cells from each other. Everything inside the cell wall was lumped together as "cell contents," "nutrient sap," "vital juice," and so on. Presently, in 1831, Robert Brown demonstrated the presence of a nucleus at least in many cells, and, in 1838, Schleiden described the formation of cells in the embryo-sac of the Phanerogam as a sort of precipitation of "gum" round definite centres. This "gum" was next shown by Naegeli to be nitrogenous, and hence wrongly named from the chemical standpoint.

Meanwhile in the animal world, where the cell wall was not nearly so prominent a feature, it had been recognised that the substance of the cell was in the main also nitrogenous, and to this substance Dujardin gave the name of "sarcode." Since the sarcode, according to the zoologists, appeared to be the active member and real basis of the cell, it began to dawn upon the botanists that the slimy contents of the vegetable cell, so like the sarcode in general appearance, might turn out to be something of the same nature, if not identical with it. Von Mohl, in 1844, took the plunge and boldly announced that the slimy lining—" primordial utricle " he called it—of the vegetable cell was a living substance and was the primary and most important constituent of the cell, and gave it the name of " protoplasm," a term he borrowed from the human physiologist Purkinje, who had so designated the granular substance of the animal ovum in 1840. The final step was not a difficult one, though it is somewhat doubtful to whom belongs the credit of having first taken it. Perhaps it would be fairest to say that the conception of the identity of sarcode and protoplasm was in the minds of several workers during the years 1845–55, but more especially in those of Max Schultze and Unger. When motile gonidia were found to be destitute of cell walls, and when De Bary had shown that cell walls were absent from the Mycetozoa, it became at once apparent that the wall was quite a subordinate feature of the cell and that the real cell consisted of a mass of protoplasm with a nucleus, an " energid," as it was afterwards termed, a view now held by all biologists.

Here then at last was the " vehicle of irritation " dimly hinted at by Knight, the real seat of the " vital force " believed in so firmly by the physiologists of the closing years of the eighteenth century, the " physical basis of life," as Huxley called it many years afterwards, and all the associated structures such as cell walls, vacuoles, oil drops, starch grains, microsomata, and so on, at once fell into their appropriate places as subordinate products of the activity of protoplasm. The so-called " vital sap " was thus seen to be protoplasm, and the movements in it were correlated by Alexander Braun and De Bary with the locomotory movements of free zoogonidia and the amoeboid movements of Mycetozoa. The stream of research was now directed to the discovery of the structure and constitution of protoplasm. After the recognition of the identity of the sarcode of the animal with the protoplasm of the plant, and after the adoption of the name "protoplasm" for both substances, botanists had the invaluable assistance of their colleagues in the sister science in unravelling the problems concerned with it that suggested themselves for solution.

Another most important question that occupied the minds of biologists at this period was the mode of origin of the varied structural units composing the plant and animal body. Research on this subject was undertaken both by zoologists and botanists, with the result that Schwann, on the animal side, and Schleiden, on the plant side, formulated independently a theory, that all the tissue elements, no matter what were their ultimate forms and functions, were derived from primary isodiametric cells, and that every cell arose from a pre-existing cell and in the long run from the fertilised ovum.

The mention of Schleiden's name in relation to the foundation of the cell theory recalls another great service that this distinguished botanist rendered to his science. Previous to 1840 the botanical student had no general textbooks to refer to save the *Théorie élémentaire de la botanique* and the *Physiologie végétale* of De Candolle. These and other treatises were replaced by the famous *Grundzüge der wissenschaftlichen Botanik* of Schleiden, published in 1842, and translated into English by Lankester, in 1849, under the title, *Principles of Scientific Botany, or Botany as an Inductive Science.* After an Introduction dealing with general principles, Schleiden devotes his first "Book" to an account of the inorganic and organic constituents of plants. The second "Book" treats of plant cells and tissues and their functions, while in the third he discusses the general and special morphology of plants. The fourth "Book" deals with

Organology or "the doctrine of the life of the whole plant and of its particular organs." These successive "Books" are well worth perusal even at the present day, not only as giving us an insight into the condition of botanical knowledge eighty years ago, but also as providing us with a résumé of the views of many authors precedent to or contemporary with Schleiden, and with the author's criticisms of their achievements—criticisms often couched in the most scathing and uncompromisingly hostile phraseology. I will quote to you a few sentences from Schleiden's fourth "Book," which I think will explain exactly the general attitude he took up with regard to both older and contemporary work, an attitude whose bluntly critical and fearlessly antagonistic expression created quite an insurrection among the botanists of his day.

"If we consider the attempts that have hitherto been made to subject the life of plants to scientific observation, we shall find that all those who have conducted them have brought to their works groundless prejudices, and, following the old beaten track, have not even paused to inquire whether or not it were right, and whether or not their prejudices were just; and they have even taken these latter as leading maxims to form the basis of all their investigations. I have already discussed the fanciful analogy between the physiology of animals and of plants. In consequence of the use of this absurd analogy, almost all the works which have hitherto appeared on vegetable physiology are perfectly worthless, for in no instance have they adopted the only true fundamental position, namely, the essential peculiarity of vegetable life; nay, the larger number of writers have not even given a comprehensive view of the facts already known, as such would have destroyed their assumed principles. Each branch of natural science, if it would lay claim to such name, must have its own peculiar independent principle of development, which must be drawn from its own data, and *only* thence. It is not until considerable advance has been made towards perfection that it is safe to begin to inquire whether analogies exist between itself and some other branch of natural science, and, if so, what they are. The manner in which science is usually pursued is not following it out gradually through a long course of original investigations, but by grasping hastily at all statements and dogmas that are afloat respecting it, seeking to participate in its treasures as an inheritance from strangers, rather than by examining into its foundations and building up its structure; this is the reason that we find even more dangerous prejudices to combat in science than in practical life. . . . Thus it has been with Botany: books have been written when plants should have been examined, conjectures have been made when investigations should have been pursued. Hence for about a century we have but revolved in a circle, without making the least advance or discovering new facts; and new laws are given us which are only the result of the play of chances, whilst correct fundamental maxims and correct methods of advance would have guaranteed the solution of various problems, and secured the progress of the science."

This sweeping condemnation of all the work of the previous century savours rather too much of Gratiano's "Sir Oracle," for remember these sentences were written scarcely fifty years after Ingen-Housz wrote his *Experiments*, and considerably less than fifty after the days of De Saussure and Knight, and surely it cannot be said that these men wrote books instead of examining plants or discovered no new facts. Of course, as was only to be anticipated, the *Principles* contained very many blunders of observation, as well as numerous quite erroneous deductions from facts observed by Schleiden himself. His criticisms also were often not only offensive in their phraseology but ridiculous in their substance. Yet with all its defects there can be no doubt that

Schleiden's textbook forms an important landmark in the history of botanical science, for it was the means of inducing workers on the subject to rub their eyes, so to speak, and reconsider the work they had accomplished as well as the researches they had in contemplation.

It is a little difficult to disentangle the various views on plant anatomy and histology that were published between 1830 and 1850, but two or three names stand out prominently, and perhaps the simplest way would be for me to follow the work of these men in succession and endeavour to estimate what additions they individually made to a knowledge of the subject.

I must first make a brief reference to Meyen, who, in 1830, wrote a *Lehrbuch der Phytotomie*. The views expressed in this book compare very unfavourably with those of Moldenhawer which I have already discussed with you, and which were published nearly twenty years before Meyen's textbook saw the light. Here we meet again with many of the blunders made by the trio that competed for the Göttingen Prize, and there are many new ones added, belonging particularly to Meyen himself. He thought that a spiral vessel was formed by first laying down a spiral thread and then secreting a wall round it afterwards. From such a vessel all the other types—annular, reticulate, and pitted—were derived, but while the spiral vessels transported sap their derivatives carried air! The laticiferous tubes are the highest type of vessel and carried the "life sap" which circulated through the plant like blood. It does not seem to have occurred to these believers in latex as "life sap" that it was necessary to explain how so many living plants had no "life sap." Then again Meyen thought stomata were cuticular glands, and that pits on the walls of cells, fibres, etc., were not depressions but elevations! After these examples of Meyen's work I do not think you will expect me to spend more time over him; you may safely class him along with Link, Rudolphi, and Treviranus.

The period of which I am speaking, viz., from 1830 to 1850, was one of great activity in anatomical and histological research, but, among the numerous workers whose publications we meet with during these years, two stand out head and shoulders above all the rest, Von Mohl and Naegeli. As we have already seen, these two men took a prominent part in establishing the great generalisation known as the " cell theory " and in the discovery of protoplasm. Over and above this, however, their contributions to general histology were important and varied, and to some of these I must now refer.

Von Mohl's earliest work was concerned with the development of tissue elements from cells, showing how the primary wall becomes modified both in external form and also as a result of secondary thickening in various ways, contradicting Meyen's view that annular, reticulate, and other types of vessels were derived from spiral. Von Mohl demonstrated also that pits were thinner places on the wall between secondary thickenings, and neither elevations nor pores. An essay on the *Anatomy of Palms* did much to clear up the architecture of the Monocotyledonous stem, and, in a paper published in 1838, we obtain at last a correct account of the structure and mode of development of stomata. A few years later Von Mohl took up Moldenhawer's interpretation of the vascular system and confirmed his idea that the vascular bundle was a compound structure composed of xylem and phloem. He traced the course of the bundles in the stem both of Monocotyledons and of Dicotyledons, and showed how the first bundles of the stem were bundles derived from the leaves—" leaf traces " as we term them, after Hanstein. He also studied the structure of the epidermis, and demonstrated the real nature of the cuticle, which had often in previous years been confused with the epidermis itself. He then turned his attention to the cork and worked out its origin and development, pointing out how, by successively deeper formations of phellogen, bark was produced, and how the isolated portions of the cortex scaled off from time to time. His best known work is, perhaps, *The Vegetable Cell*, first published in 1851, and translated at a later date into English. In this volume he summarises the principal work that had been carried out on the cell by himself and by others, and for many years it formed the reference textbook on all questions connected with the subject of cytology as it was then understood.

The work of Payen must not be passed over without notice. To him we owe our earliest knowledge of the chemistry of the cell wall, for he showed that every cell wall was at first composed of cellulose, and that as age increased the cellulose might become altered by the addition of "incrusting bodies" which, however, were removable by the aid of certain chemical reagents, thus paving the way for the broader generalisations of cutination, suberisation, mineralisation, lignification, and so on, that we meet with at a later date.

Naegeli began his work by an investigation of the Thallophyta and was among the first to study the morphology and anatomy of Algae with a view to their natural classification. He did a considerable amount of work on the growing apex of the stem and root, and traced the development of all permanent tissues to the segments of an apical cell in the lower plants and of an apical meristem in the higher forms. He also confirmed Unger's views on cell division in these apical regions, and exposed the fallacy of "free cell formation" as advanced by Schleiden.

During the years 1847–49 Naegeli worked out the taxonomy of some groups of Algae, and made many additions to our knowledge of their structure and life history. He studied the development of vascular bundles from the procambial stage and, together with Hanstein and Sanio, succeeded in explaining the mode of secondary thickening in the stems of Dicotyledons, a point in which

Von Mohl had gone astray. He also elucidated the structure of sieve tubes which had been identified by Hartig in 1851. He formulated a classification of tissues, distinguishing them into generative and permanent, both being again subdivided into parenchymatous and prosenchymatous types. The primary parenchymatous generating tissue occurred, according to Naegeli, in all embryos and young organs, while the prosenchymatous meristem, which he called "cambium," was arranged in strands or layers in fully developed organs and between permanent tissues. The permanent tissues thus took their origin either in primary or secondary meristem. Accepting Hanstein's views as to leaf trace bundles, he further distinguished regions in the bundles as foliar, common, and cauline.

Even from this brief summary of the researches of these two men you will appreciate how much plant anatomy owes to them; but their labours were not confined to microscopic investigations, for we shall find their names cropping up again and again during the next decade or two in connection with other developments of the science. Meanwhile you will observe that histology has begun to wear a more familiar aspect, and to resemble in many respects the accounts met with in present-day treatises on the subject. There are still many details to be filled in, but the foundations are securely laid.

LECTURE VI

HOFMEISTER ON THE VASCULAR CRYPTOGAMS

WHILE Von Mohl and Naegeli were busily engaged in working out the real nature of the tissues, more especially of flowering plants, another and even greater anatomist, Wilhelm Hofmeister, was producing a series of monographs on the Archegoniatae which were destined to make his name famous for all time. Not only did his researches throw a flood of light on the structure and life histories of the Pteridophyta and Gymnosperms but, by the facts they revealed, they made it possible to see a unity of plan throughout the vegetable kingdom, and broke down the barrier that had hitherto existed between the lower and higher plants; they permitted botanists to see the plant world as a whole, and to apply to it the phylogenetic principles which were on the eve of being given to the world by Charles Darwin.

Hofmeister's work began in reality in 1840 with a paper on the origin of the embryo in the Phanerogams, wherein he showed that an ovum was present in the embryo-sac before fertilisation, and that, after fusion with the contents of the pollen tube, an embryo was formed from it. He traced the formation of this embryo from the oosperm up to the resting condition in the seed, and thus confirmed and completed the earlier work of Amici and Robert Brown. This, however, was only the starting point of his great undertaking. He proceeded next to investigate the life cycles of representatives of the Bryophyta and Vascular Cryptogams and, later, of the

Gymnosperms also, and demonstrated that in all of them there was an alternation of a sexual with an asexual generation, each starting from a single cell. These results were published between 1849 and 1851 under the title *Vergleichende Untersuchungen höherer Kryptogamen,* a treatise which was translated into English in 1862, at the instance of the Ray Society, and also in 1852 as *Beiträge zur Kenntis der Gefässkryptogamen,* etc., while several genera were dealt with individually at later dates, such as Riella in 1854 and Salvinia in 1857.

It is quite impossible to give you any adequate idea of the contents of these and other researches published by Hofmeister during the decade 1849–59, but when I tell you that in them we are presented with detailed accounts of the anatomy and life cycles of many Bryophyta and also of the chief genera of Pteridophyta, such as Pilularia, Salvinia, Isoetes, Ophioglossum, Equisetum, and Selaginella, together with comparisons of these and other genera with Coniferae and Angiosperms, you will readily recognise that Hofmeister's work was, in the true sense of the term, epoch-making. As I have already said, Hofmeister showed that alternation of a sexual with an asexual generation was common to all the plants usually spoken of as Mosses and Vascular Cryptogams, and further that these two phases could be recognised in Gymnosperms and Angiosperms as well. In the Ferns the spore gave rise to an inconspicuous, green, parenchymatous plant, the prothallus, on which archegonia and antheridia were produced, and the fertilised ovum in turn developed into the fern plant, or sporophyte, producing spores once more and thus completing the cycle. In Mosses the spore developed into a much more highly differentiated gametophyte with stem and leaves, or structures that at least corresponded to these organs ; but in this case the fertilised ovum did not become an independent plant but a semi-parasitic sporocarp or sporogonium, whence in turn spores were derived.

Hofmeister next showed that in Pilularia, Salvinia, Isoetes, and Selaginella there were two kinds of spore, a smaller one giving rise to a male and a larger one to a female gametophyte, the fertilised ovum of the latter becoming once more the generation bearing the generic name. In other words he elucidated the true significance of heterospory, and followed out the differentiation of sexuality in the gametophytes of the Pteridophyta. Turning next to the Coniferae, he demonstrated that the pollen grain in that group was the equivalent of the microspore in the heterosporous Vascular Cryptogams, while the embryo-sac and endosperm corresponded to the megaspore and its contained female prothallus. The homologies between the two stages in the Vascular Cryptogams and the condition of affairs in flowering plants now presented no difficulties, and the ovule was seen to consist of a single megaspore permanently enclosed within its sporangium, whose wall had become greatly thickened and nutritive in function—Robert Brown's "perisperm"—and had acquired additional integuments, the whole enclosed in the carpel or modified sporophyll.

The chief effect of all these fundamental discoveries was to show that the key to a true natural classification lay not in the balancing of values among the various morphological parts of the flower and fruit, but in a detailed study of the anatomy of the real and concealed reproductive organs, rather than of the more obvious pseudo-organs of multiplication. How great was the change in the whole outlook we shall see later on; meanwhile the splendid results achieved by Hofmeister were the means of turning the attention of a host of investigators to the group that had yielded such a harvest of illuminating ideas, the hitherto despised and neglected Cryptogams.

Progress in Physiological Problems

Leaving these important investigations for the present, let us collect the various threads of research in other departments, more especially in physiology, as we left them in a previous lecture.

While Liebig and Boussingault were replacing the "humus theory" with sounder views on the relationship of the soil to the plant, other workers had been trying to solve the even more difficult problem of photosynthesis or, as it was then called, carbon assimilation. As far back as 1819 Peletier and Caventou had introduced the term "chlorophyll" to indicate the green pigment in plants, as to the significance of which so little was as yet known. The influence of sunlight on the gaseous exchange between the plant and the atmosphere had been studied by the investigators of the concluding years of the eighteenth century and by De Saussure in the beginning of the nineteenth, but in 1834 the chemist Dumas tried to carry the matter a step farther by attempting to differentiate between the effects of the different rays of which white light is composed, and hazarded the suggestion that the more refrangible blue rays—as being those most markedly absorbed by chlorophyll—were those chiefly concerned in the photosynthetic process, an unfortunate guess as it turned out. This new line of enquiry was the direct outcome of Brewster's discovery that an alcoholic solution of chlorophyll gave a very characteristic absorption spectrum, and this fact, coupled with Dumas' speculation, led Daubeny, in 1836, to experiment on plants grown behind coloured glass screens. He found that little or no assimilation took place in plants exposed to blue and violet light, and expressed the opinion that the most efficacious rays were those in the yellow region of the spectrum. About the same time Dutrochet introduced the method, so frequently employed in later years, of determining the activity of carbon dioxide decomposition under different conditions by counting and analysing the gas bubbles given off by submerged plants in sunlight. In 1838 Von Mohl made the first serious attempt to elucidate the nature of the chlorophyll apparatus. He showed that the pigment did not occur in solution in the cell but existed in the form of minute green viscid masses of varied shape, and proved the presence in them of granules of starch by treating the tissue with iodine solution. It should be remembered that Grew spoke of chlorophyll as a "green precipitate," so that the solid condition in which the pigment existed in the cell was not altogether a new discovery. Further, the "gas bubble" method of determining the activity of carbon dioxide decomposition first employed by Dutrochet is usually credited to Sachs, as is also the identification of starch granules in chloroplasts by the iodine test. It is remarkable that in the prominent textbook of the period, the *Grundzüge* of Schleiden, all the work of Ingen-Housz, De Saussure, Dutrochet, and others I have mentioned, is completely ignored or flatly contradicted. On this particular question of carbon-assimilation Schleiden shows himself retrogressive in the extreme. "Thus it appears," he writes, "at least as far as the facts carry us, to be completely proved that plants are not nourished by the carbonic acid of the atmosphere by aid of their green parts" (auf diese Weise scheint, wenigstens soweit bis jetzt die Thatsachen vorliegen, vollkommen festzustehen, dass die Pflanzen sich nicht auf Kosten der Kohlensäure der Atmosphaere durch die grünen Theile nähren).

While Daubeny used coloured glass screens in his experiments Draper employed the spectrum itself. He placed leaves of various plants in water impregnated with carbon dioxide and submitted them to the rays from different parts of the spectrum and estimated the activity of decomposition by measuring the amounts of oxygen given off, with the aid of an eudiometer. The results he obtained showed that the maximum activity took place in the brightest region of the spectrum; no decomposition of carbon dioxide occurred under the more refrangible rays nor under the dark heat rays.

Another investigator, Garreau, did considerable service in 1849 and 1851 by determining the relative values of the upper and under surface of the leaf as paths of gaseous exchange, more especially directing his attention to the exhalation of water vapour. In 1838 Berzelius and also, in 1851, Verdeil made attempts at isolating pure chlorophyll. The latter treated a hot alcoholic extract of leaves with lime water and then added strong hydrochloric acid. The research in itself was of little or no importance, but I mention it as many experimenters from that day onwards strove to extract the pigment in a pure condition by the use of reagents of the most drastic character, ignoring the likelihood of the manufacture thereby of decomposition products from so complex and unstable a substance during extraction and purification.

The only other paper in relation to chlorophyll I need refer to is one by Gris who, in 1857, showed that seeds which contained little or no iron, if grown in a medium free from that metal, produce only a few partially green leaves, and that subsequent leaves are quite chlorotic, but that if a few drops of a solution of an iron salt be added to the culture and if transpiration be active, these chlorotic leaves become green in a few days. This result gave rise to the belief that iron was an essential constituent of chlorophyll, a view, however, that was afterwards found to be incorrect.

The Foundations of Palaeophytology

There is one other department of botanical research to which I have not as yet referred at all, viz., palaeobotany. Although the study of fossil plants is of comparatively recent growth, the classical writers were not altogether ignorant of the existence of fossils. These were, however, remains of animal life and there is no sure evidence that they identified any petrifactions as of vegetable origin. When fossils really began to be collected and studied in the end of the sixteenth and early years of the seventeenth centuries they were looked upon as the remnants of vegetable and animal life that had existed on the earth previous to the Noachian Deluge, and it is somewhat amusing to read in Luther's commentary on Genesis the view expressed that indications of the Deluge would be found in the form of petrified wood near mines and other centres of early man's activities!

Johann Scheuchzer brought out a great work in 1709 on the supposed remains of the Deluge, one of which was the famous Salamander fossil which he called "Homo diluvii testis," and which he believed to be the skeleton of a man who had witnessed the flood! By the end of the eighteenth century, however, the Deluge theory had begun to fall into disrepute, and naturalists adopted saner views on the nature of the objects that were constantly being extracted from the earth's crust. Apparently the first to enunciate sounder scientific ideas on fossils was Johann Blumenbach, who, in 1790, asserted that the organisms of which these fossils were the remains were pre-Adamitic, and that there had been many faunas and floras on the earth before the advent of man.

Undoubtedly the founder of modern palaeophytology is Adolphe Brongniart, who, in 1828, began the publication of his *Histoire des végétaux fossiles.* "The systematic manner in which the science was organised and built up by him made him the highest authority on the subject of fossil plants, and the numerous more or less elaborate memoirs that continued to appear showed that none of the minor details were neglected." Brongniart divided the geological series into four periods; in the first the dominant vegetation was chiefly cryptogamic, composed of Ferns, Lepidodendra, and arboreal Equiseta; in the next epoch the lower plants were replaced by forms that have no modern representatives, *e.g.* specialised Ferns and Conifers, such as we now regard as belonging to the Triassic horizon. The third period was the Mesozoic, when Cycads abounded; while the fourth period was characterised by the presence of plants not markedly different from those now existing. The *Histoire* never got beyond the first volume and a fragment of a second, dealing with the Lepidodendra, which he regarded as intimately related to the Lycopodiaceae.

Witham, a Yorkshire palaeobotanist, deserves to be mentioned as the first to study the Carboniferous plants histologically. His principal work was a paper on *The Internal Structure of Fossil Vegetables Found in the Carboniferous and Oolitic Deposits of Great Britain*, published in 1833.

Another great name in the history of fossil botany was that of Robert Goeppert. " Endowed with the true German devotion to his specialty, with keen observing and analytic powers, with a restless activity, exceptional opportunities and a long life, he was able to create for the science a vast wealth of new facts and give it a solid body of laboriously wrought truth. If Brongniart laid the foundations of palaeobotany Goeppert may properly be said to have built its superstructure." His first important work was his *Systema Filicum Fossilium*, published in 1836, in which he compared fossil and living Ferns and illustrated the former profusely in well-executed plates. In 1841 there followed the *Genera of Fossil Plants*, which appeared both in German and French, and also a monograph on the fossil flora of Silesia, Goeppert's native district.

In 1845 several important additions to the literature of palaeobotany were made, such as Unger's *Synopsis Plantarum Fossilium*, Corda's *Flora der Vorwelt*, and Goeppert's study of amber. Unger's work in this and succeeding years was specially important for his views on classification. He anticipated ideas now beginning to be held, viz., that Cycads were more closely related to Cryptogams than to Conifers, and refused to regard Stigmaria as a Dicotyledon, as had been held by English botanists. The true relationships of these and other arboreal remains from the Carboniferous rocks were not, however, clearly made out until the later years of the century, when Williamson began his epoch-making work on the *Fossil Plants of the Coal Measures.*

CHARLES DARWIN

The year 1859 was destined to be a fateful one not merely for science in general and botany in particular but for every department of learning, for in that year was published a book that has had a deeper and more wide-reaching influence on the trend of human thought and endeavour than any other that has ever come from the printing-press—I mean, of course, *The Origin of Species* by Charles Darwin. Needless to say I do not intend to ask your attention just now to more than those parts of Darwin's life-work that are primarily botanical, although it will be necessary for you to comprehend, in some measure at least, the bearing of his larger views on the science in which we are at present more particularly interested. Papers, pamphlets, books, almost libraries, one might say, have been and are still being written by authors, scientific and unscientific, dealing with Darwin's life and labours; to these I must refer you should you desire to know more than I have time to tell you, and to one work more especially, *The Life and Letters*, by his distinguished son, Sir Francis Darwin, whose name is so familiar to you as the author of numerous important monographs on various sections of plant physiology. The brief account I am about to give you is taken chiefly from these volumes and from the essays of Darwin's friend and champion, Huxley.

Charles Darwin was born in 1809 at Shrewsbury, where his father was a well-known medical man. His mother was a daughter of the founder of the famous Wedgwood Potteries. It is rather amusing to read how Darwin's early teachers looked upon him as not far removed from a dunce. These were the days when higher education was considered as synonymous with an intimate knowledge of the classics, and it is on record that his headmaster publicly reprimanded Darwin for "wasting his time on such a contemptible subject as chemistry." "Omnia mutantur, nos et mutamur in illis," and the narrow-minded classicist who regards Latin and Greek as the one and only basis of a liberal education is rapidly succumbing to the fate that has overtaken the Lepidodendron and the Ichthyosaurus. "The School as a means of education to me was simply a blank," Darwin himself writes, and Huxley adds, "As a matter of fact Darwin's school education left him ignorant of almost all the things which it would have been well for him to know, and untrained in all things it would have been useful for him to be able to do in after life. Drawing, practice in English composition, and instruction in the elements of the physical sciences would not only have been infinitely valuable to him in reference to his future career, but would have furnished the discipline suited to his faculties, whatever that career might be."

Darwin does not appear to have been much more of a success at college than he was at school. At Edinburgh, where he went to study medicine, he found the professors "intolerably dull," and says that the lectures of the professor of Materia Medica were "fearful to remember," while he describes the discourses of the professor of Anatomy as having been "as dull as he was himself." As Huxley puts it, "the climax seems to have been attained by the professor of Geology and Zoology, whose praelections were so incredibly dull that they produced in their hearer the somewhat rash determination never to read a book on geology or in any way to study the science so long as he lived." This, remember, was the decision of the future author of the *Structure and Distribution of Coral Reefs* and of the *Geological Observations on Volcanic Islands*.

Finding himself not likely to become a successful medical man, Darwin exchanged Edinburgh for Cambridge, where he entered Christ's College with the view of reading for holy orders, but he had no better words to say for Cambridge than he had for Modern Athens. "During the three years which I spent at Cambridge," he says, "my time was wasted, as far as academic studies were concerned, as completely as at Edinburgh and as at school." If he passed the door of the geology lecture room with a cold shudder, he found his way into the botanical one and there he made the acquaintance of Henslow, "a man of rare character and singularly extensive acquirement in all branches of natural history." The acquaintance grew into friendship which lasted till Henslow's death in 1861. Henslow overcame Darwin's prejudice against geology and succeeded in introducing him to Sedgwick, at that time professor of geology at Cambridge, and Darwin had to forswear his vow never to study the science, for he not only accompanied the professor on his geological excursions but set himself to master Lyell's *Principles of Geology*, a work to which in after years he professed himself as fundamentally indebted.

Henslow is, however, responsible for more than merely turning Darwin's thoughts to botany and geology; he showed himself possessed of a far-seeing vision that was the means of dedicating to science the man who was destined to become perhaps the greatest of her priests. For it was due to Henslow that Darwin, when scarcely out of his teens, was appointed naturalist on the *Beagle* for a five years' surveying voyage round the world. "So," again to quote Huxley, "a fourth educational experiment was to be tried; this time Nature took him in hand herself," and where the Shrewsbury pedagogue, the Edinburgh medicos, and the Cambridge theologians had signally failed, Nature showed him the way by which, to borrow Henslow's prophetic phrase, "anything he pleased might be done."

And Nature, the old nurse, took The child upon her knee,
Saying, "Here is a story-book Thy Father has written for thee."

It is both interesting and amusing to think that these lines were written by Longfellow on Agassiz, an uncompromising opponent of the Evolution Theory.

This fourth educational experiment was as successful as the others had been fruitless, and was, as Darwin himself puts it, the starting-point of his "second life."

The only item connected with the voyage of the *Beagle* I propose to refer to occurs in an extract from one of his letters written in 1877. "When I was on board the *Beagle,* I believed in the permanence of species, but, as far as I can remember, vague doubts occasionally flitted across my mind. On my return home in the autumn of 1837 I immediately began to prepare my journal for publication and then saw how many facts indicated the common descent of species, so that in July 1837, I opened a notebook to record any facts which might bear on the question. But I did not become convinced that species were mutable until, I think, two or three years had elapsed."

Soon after his return from his travels Darwin came across a famous book called *An Essay on the Principles of Population,* written by the Rev. Thomas Malthus, a Surrey vicar, who had died whilst Darwin was abroad. The essay was a very pessimistic production aimed at meeting the rhapsodic views of the Rousseau school of thinkers. Malthus argued that organisms if left perfect freedom to breed and given unlimited room in which to multiply would fill any conceivable area in a very short time. The only limits to their increase were those of space and food. So far as man was concerned, propagation was controlled by reason, although, in his case also, the same limitations were existent and active, and these he proceeded to enumerate and illustrate from the histories of different nations. "The necessary result of unrestricted multiplication," said Huxley, "is competition for the means of existence. The success of one competitor involves the failure of the rest, that is, their extinction; and this selection is dependent on the better adaptation of the successful competitors to the conditions of the competition. Variation occurs under natural no less than under artificial conditions. Unrestricted multiplication implies the competition of varieties and the selection of these which are relatively best adapted to the conditions. An individual which varies, *ipso facto* diverges from the type of its species; and its progeny, in which the variation becomes intensified by selection, must diverge still more, not only from the parent stock, but from any other race of that stock starting from a variation of a different character. The selective process could not take place unless the selected variety was either better adapted to the conditions than the original stock, or adapted to other conditions than the original stock. In the first case, the original stock would be sooner or later extirpated; in the second, the type, as represented by the original stock and the variety, would occupy more diversified stations than it did before."

Darwin wrote out his theory in 1844, but instead of publishing it at once he spent the next fifteen years gathering new data bearing on the various aspects of the subject, conducting experiments, and, so to speak, polishing the rough-hewn marble into the perfect statue. During this period also Darwin was busily engaged in working out other problems suggested by or connected with his main thesis, the gist of which latter was known only to a few intimate friends. At length in 1856, at Lyell's instigation, Darwin began to expand the preliminary sketch of his theory that he had written twelve years before so as to bring it into book form, but with no immediate intention of publishing. In May 1857 he wrote: "I find the subject so very large, that, though I have written many chapters, I do not suppose I shall go to press for two years."

Then there descended a bolt from the blue that forced his hand. Alfred Russel Wallace, a young architect, who had for some years exchanged his practice for a life of travel and exploration in Brazil and Malaya, and with whom Darwin had had considerable correspondence on natural history subjects, sent him an essay on *The Tendency of Varieties to Depart Indefinitely from the Original Type*, the outcome of his observations and meditations amid the tropical forests of the Eastern Archipelago, and which, as Darwin says, was, in effect, an admirable abstract of his own unpublished book. After consultation with his two most intimate friends Lyell and Hooker, he decided to publish Wallace's essay and an abstract of his own work simultaneously, and the two papers were thus read to the Linnean Society on July 1st, 1858, a truly historic date and one to be remembered by every student of biology.

A little over a year thereafter appeared the famous volume, *On the Origin of Species by Means of Natural Selection, or the Preservation of Favoured Races in the Struggle for Life.* I am not going to attempt to summarise in my own words a work of which Hooker writes, "It is the very hardest book to read to full profit, that I ever tried," and which Huxley, one of the clearest-headed of the biologists of the last century, pronounced "one of the hardest books to master." I prefer that you should have Huxley's own summary as given in his obituary notice of Darwin to the Royal Society, and also Wallace's presentation of the theory published in his essay on "Creation by Law," in the *Quarterly Journal of Science* in 1868.

Huxley's résumé is as follows: "Observation proves the existence among all living beings of phenomena of three kinds, denoted by the terms heredity, variation, and multiplication. Progeny tend to resemble their parents; nevertheless all their organs and functions are susceptible of departing more or less from the average parental character; and their number is in excess of that of their parents. Severe competition for the means of living, or the struggle for existence, is a necessary consequence of unlimited multiplication; while selection, or the preservation of favourable variations and the extinction of others, is a necessary consequence of severe competition. 'Favourable variations' are those which are better adapted to surrounding conditions. It follows, therefore, that every variety which is selected into a species is so favoured and preserved in consequence of being, in some one or more respects, better adapted to its surroundings than its rivals. In other words, every species which exists, exists in virtue of adaptation, and whatever accounts for that adaptation accounts for the existence of species."

Wallace's epitome is in the form of a table consisting of two parallel columns, the first containing proved facts; the second, legitimate deductions from these facts, which deductions are afterwards transferred to the first column as proved facts. The table has at least the merit of placing vividly before you the fundamental principles on which the whole theory rests, although you must bear in mind that each sentence really stands for a volume. It is more in the nature of a guide to a library than a key to any one volume in it. Starting with the two acknowledged facts, (1) that organisms left to themselves increase at an enormously rapid rate, and (2) that, despite this, the total number of individuals at any given time remains on the whole stationary, the conclusion drawn is, (3) that there is a struggle for existence, the deaths on the average equalling the births. Accepting the struggle for existence as a proved fact and adding to it, (4) the facts of heredity and variation, or the general likeness, with individual differences, of parents and offspring, we reach the conclusion, (5) that on the whole those die who are least fitted to maintain their existence—survival of the fittest or natural selection. Again accepting this conclusion as a fact and taking it in conjunction with, (6) the fact that the changes in external conditions are universal and unceasing, the deduction is, (7) that changes of organic forms must take place to keep them in harmony with the changed conditions; and as the changes of conditions are permanent changes, in the sense of not reverting back to identical previous conditions, the changes of organic forms must be in the same sense permanent and thus originate new species.

"It is doubtful," writes Huxley, "if any single book, except the *Principia* [of Newton] ever worked so great and so rapid an evolution in science, or made so deep an impression on the general mind. It aroused a tempest of opposition and met with equally vehement support, and it must be added that no book had been more widely and persistently misunderstood by both friends and foes."

During his voyage in the *Beagle* Darwin suffered much from the effects of a serious illness contracted at Valparaiso, and this left its mark upon him to such an extent that, as he tells us in his autobiography, he never really recovered his initial health and strength. "My chief enjoyment and sole employment throughout life," he writes, "has been scientific work, and the excitement from such work makes me, for the time, forget, or drives quite away, my daily discomfort. I have, therefore, nothing to record during the rest of my life, except the publication of my several books." After 1842 Darwin removed from London to a country residence at Down in Kent, where he spent the remainder of his years.

The first of the books he refers to in the sentence

I have just quoted was *On the Various Contrivances by which Orchids are Fertilised by Insects*, which was really an appendix to Sprengel's work, *The Secret of Nature Revealed*, published in 1793, about which I spoke to you recently, and aimed at supplying the missing key to Sprengel's observation, viz. "that crossing played an important part in keeping specific forms constant." Darwin came to the further conclusion that "cross fertilisation [or cross pollination, as we would call it nowadays] is favourable to the fertility of the parent and to the vigour of the offspring," and that "all those mechanisms which hinder self-fertilisation and favour crossing must be advantageous in the struggle for existence; and that the more perfect the action of the mechanism the greater the advantage."

This preliminary effort, which was published in 1862, led Darwin to undertake a more extended enquiry into the subject, under the title, *The Effects of Cross- and Self-Fertilisation in the Vegetable Kingdom*, which did not appear, however, until 1876. In the following year another of his books was published dealing with *The Different Forms of Flowers on Plants of the Same Species*, in which he discussed the significance of the di- and trimorphic flowers of Primula, Lythrum, Linum, Oxalis, Viola, and other genera which showed this peculiarity.

"In the course of the twenty years during which Darwin was thus occupied in opening up new regions of investigation to the botanist and showing the profound physiological significance of the apparently meaningless diversities of floral structure, his attention was keenly alive to any other interesting phenomena of plant life which came in his way. In his correspondence he not infrequently laughs at himself for his ignorance of systematic botany; and his acquaintance with vegetable anatomy and physiology was of the slenderest. Nevertheless, if any of the less common features of plant life came under his notice, that imperious necessity of seeking for causes which nature had laid upon him, impelled, and indeed compelled, him to inquire the how and the why of the fact, and its bearing on his general views. And as, happily, the atavic tendency to frame hypotheses was accompanied by an equally strong need to test them by well-devised experiments, and to acquire all possible information before publishing his results, the effect was that he touched no topic without elucidating it.

"Thus the investigation of the operations of insectivorous plants, embodied in the work on that topic, published in 1875, was started fifteen years before, by a passing observation during one of Darwin's rare holidays. 'In the summer of 1860 I was idling and resting near Hartfield, where two species of Drosera abound; and I noticed that numerous insects had been entrapped by the leaves. I carried home some plants and on giving them some insects saw the movements of the tentacles, and this made me think it possible that the insects were caught for some special purpose. Fortunately, a crucial test occurred to me, that of placing a large number of leaves in various nitrogenous and non-nitrogenous fluids of equal density; and as soon as I found that the former alone excited energetic movements, it was obvious that here was a fine new field for investigation.'

"The researches then initiated led to the proof that plants are capable of secreting a digestive fluid like that of animals, and of profiting by the result of digestion; whereby the peculiar apparatuses of the insectivorous plants were brought within the scope of natural selection. Moreover, these inquiries widely enlarged our knowledge of the manner in which stimuli are transmitted in plants, and opened up a prospect of drawing closer the analogies between the motor processes of plants and those of animals.

"So with respect to the books on *Climbing Plants* (1875), and on the *Power of Movement in Plants* (1880), Darwin says: 'I was led to take up this subject by reading a short paper by Asa Gray, published in 1858. He sent me some seeds, and on raising some plants I was so much fascinated and perplexed by the revolving movements of the tendrils and stems, which movements are really very simple, though appearing at first sight very complex, that I procured various other kinds of climbing plants and studied the whole subject. . . . Some of the adaptations displayed by climbing plants are as beautiful as those of orchids for ensuring cross fertilisation.'

" In the midst of all this amount of work, remarkable alike for its variety and its importance, among plants, the animal kingdom was by no means neglected. A large moiety of *The Variation of Animals and Plants Under Domestication* (1868), which contains *pièces justificatives* of the first chapter of the *Origin*, is devoted to domestic animals, and the hypothesis of ' pangenesis ' propounded in the second volume applies to the whole living world. In the *Origin* Darwin throws out some suggestions as to the causes of variation, but he takes heredity, as it is manifested by individual organisms, for granted, as an ultimate fact ; pangenesis is an attempt to account for the phenomena of heredity in the organism, on the assumption that the physiological units of which the organism is composed give off gemmules, which in virtue of heredity, tend to reproduce the unit from which they are derived." (Huxley.)

Darwin's subsequent works on such subjects as the *Descent of Man*, the *Expression of the Emotions*, etc., do not concern us here, though the former book created well-nigh as great a sensation as the *Origin* itself.

In the beginning of 1882 Darwin's health, always feeble, began to give way very rapidly, and he breathed his last on April 19th of that year, at the age of seventy-three.

As I would like you to carry away with you from these lectures not only some conception of the achievements of the leaders in botanical thought but also a mental picture, however sketchy, of their personalities, I make no apology for quoting the closing sentences of Huxley's noble panegyric, which appeared a week later in the pages of *Nature*:

"It is not for us to allude to the sacred sorrows of the bereaved home at Down; but it is no secret that, outside that domestic group, there are many to whom Mr. Darwin's death is a wholly irreparable loss. And this not merely because of his wonderfully genial, simple, and generous nature; his cheerful and animated conversation, and the infinite variety and accuracy of his information; but because the more one knew of him, the more he seemed the incorporated ideal of a man of science. Acute as were his reasoning powers, vast as was his knowledge, marvellous as was his tenacious industry, under physical difficulties which would have converted nine men out of ten into aimless invalids; it was not these qualities, great as they were, which impressed those who were admitted to his intimacy, with involuntary veneration, but a certain intense and almost passionate honesty by which all his thoughts and actions were irradiated, as by a central fire.

"It was this rarest and greatest of endowments which kept his vivid imagination and great speculative powers within due bounds, which compelled him to undertake the prodigious labours of original investigation and of reading, upon which his published works are based; which made him accept criticisms and suggestions from anybody and everybody, not only without impatience, but with expressions of gratitude sometimes almost comically in excess of their value; which led him to allow neither himself nor others to be deceived by phrases, and to spare neither time nor pains in order to obtain clear and distinct ideas upon every topic with which he occupied himself.

"One could not converse with Darwin without being reminded of Socrates. There was the same desire to find some one wiser than himself; the same belief in the sovereignty of reason; the same ready humour; the same sympathetic interest in all the ways and works of men. But instead of turning away from the problems of Nature as hopelessly insoluble, our modern philosopher devoted his whole life to attacking them in the spirit of Heraclitus and of Democritus, with results which are the substance of which their speculations were anticipatory shadows.

"The due appreciation, or even enumeration, of these results is neither practicable nor desirable at this moment. There is a time for all things—a time for glorying in our ever extending conquests over the realm of Nature, and a time for mourning over the heroes who have led us to victory.

"None have fought better, and none have been more fortunate, than Charles Darwin. He found a great truth trodden under foot, reviled by bigots, and ridiculed by all the world; he lived long enough to see it, chiefly by his own efforts, irrefragably established in science, inseparably incorporated with the common thoughts of men, and only hated and feared by those who would revile, but dare not. What shall a man desire more than this? Once more the image of Socrates rises unbidden, and the noble peroration of the 'Apology' rings in our ears as if it were Charles Darwin's farewell:

> The hour of departure has arrived, and we go our ways—
> I to die and you to live. Which is the better, God only knows."

LECTURE VII

THE INFLUENCE OF THE "ORIGIN OF SPECIES" ON BOTANICAL RESEARCH

THE *Origin of Species* appeared most opportunely, for it supplied the key required to elucidate the wonderful discoveries that had been made by Hofmeister. As I have already told you, his investigations led to the recognition of a uniformity in life-history throughout the whole plant world, and at last united in one continuous series the Cryptogams and Phanerogams. The key was progressive evolution: the result was a phylogenetic classification based on alternation of generations. The vast importance of physiology now became evident in the light of adaptation to changes of environment. Living plants had now to be pictured as a host of units in competition with each other for food and for space in which to live and multiply, fighting for existence, on the one hand against each other, even more especially against their own kith and kin, and on the other against unfavourable conditions of the environment. Every structural detail had to be explained in terms of vital needs, and morphology and its extension, anatomy, were seen to be the concrete and visible manifestations of physiological adaptation.

The dogma of the constancy of species, that had been an incubus on the shoulders of all the taxonomists of past generations, and that had driven them to all sorts of subterfuges in their efforts to reconcile self-evident facts with preconceived ideas, at last begins to disappear and the plant world is revealed as a great tree, its roots buried among the primeval unicellular forms of the remotest past, the tips of the loftiest twigs the living plants of the present, some tracing their descent almost from the original stock, others springing out at different levels from the main trunk or from its lateral branches, others still, and these the most prominent though at the same time the most recent, arising in dense tufts from the topmost branches, like the luxuriant outgrowths at the apex of a pollarded willow. Here then was opened up for the taxonomist a field of labour of boundless extent, but the unravelling of the maze was not a task that could be accomplished at once or by any single man.

Let me attempt, first of all, to give you a summary of the general situation in botanical science just after the publication of the *Origin* and before the new leaven had begun to make its influence felt on the various departments into which our science had now become divided.

For many years after 1859 we meet with attempts to sketch out the phylogeny of special groups, but no comprehensive effort to produce an entirely new taxonomic scheme on genealogical lines. Up to 1860 there was nothing better available than the Candollean system, which indeed maintained its ascendancy for over half a century more.

In morphology the views of Wolff, who regarded "all parts of the plant except the stem as modified leaves," and of Goethe, as extended by Alexander Braun, who considered all leaves as modifications of an ideal leaf, were not productive of any very clear insight into the architecture of the plant as a whole. The conception of the flower was based on the idea of a progressive metamorphosis of the appendages of the floral axis from the foliage leaf through the bract to the sepal, petal, stamen, and carpel, so that the flower was regarded as a greatly modified vegetative shoot.

In anatomy the investigations, more especially, of Schleiden, Von Mohl, and Naegeli, had led to the recognition of the cell as the fundamental unit in plant construction, and protoplasm as the physical basis of all plant activities. It had been established that all the varied types of cell, fibre, and vessel that went to the formation of the different organs were derived from homogeneous primary cells at growing points at the ends of roots and shoots, or from layers or strands of similar embryonic cells interpolated between permanent tissues, but the knowledge that had been acquired of the mode of differentiation of the permanent tissue elements from the initial embryonic cells, and the relationship of the different types of tissues to each other, was still very fragmentary. What was as yet awanting was the vivifying influence of the conception of physiological division of labour, and that conception was necessarily bound up with the idea of adaptation of structure to function.

In physiology also matters were in a backward state; no physiologist of any particular note had succeeded Dutrochet. Liebig and Boussingault had certainly done much to clear up the question of the relationship of the plant to the soil, more especially by upsetting the humus theory of nutrition, and by determining the true source of nitrogen, but both these men were chemists rather than botanists. The problems of the ascent of sap, as also of the precise function of chlorophyll in nutrition, were still unsolved, and botanists were not even yet perfectly clear as to the universal occurrence of respiration in plants.

That all plants, save the very lowest, possessed sexuality had just been realised, although in many groups the reproductive organs were still unknown, and it was only when the universal occurrence of protoplasm in all living cells had been accepted as a fact, and when the peculiar properties of that substance began to be appreciated, that the physiology of sensitivity attracted attention.

Palaeophytology as an integral branch of botany had scarcely got beyond the collection of certain independent data by men like Goeppert and Brongniart, and taxonomists of living plants still ignored the existence of fossil types. The section of botany we know under the name of ecology had not of course come into being.

The influence of Darwin's great generalisation was so profound and cumulative in its effects that I find it necessary to deal with the succeeding years of this history on a plan somewhat different from that I have followed up till now. Hitherto, as you may have noticed, each epoch has been marked by the predominance of some great name, such as that of Hales, Ray, Linnaeus, or Ingen-Housz; but after 1860 we are more concerned with great ideas, in the establishment or advancement of which many investigators took part. I think therefore it will be preferable to follow out each of these ideas separately, paying more attention to the phases through which they passed than to the precise years in which the successive advances took place. After all, the half century following 1860 is almost a complete era in itself; there are no barren gaps in it; research was continuous and cumulative in every branch of the science.

Perhaps the most noticeable feature is the gradual linking together of the different departments of botany, and the mutual help given by each to the others, dominated by the central idea of the plant as a living organism exhibiting progressive differentiation of structure with physiological division of labour among its constituent parts, and leading to the grouping of the varied forms on a genuine phylogenetic basis.

Another important development was the recognition of the fact that botany as a science could not stand alone. In its more strictly biological aspect it linked itself on to zoology, and on its physiological side it demanded and received the aid of physics and chemistry, while geology was put under contribution not merely to determine the horizons when certain extinct forms appeared and disappeared, but also, in later epochs, to explain anomalies in the distribution of living types.

Alternation of Generations

The central idea of Hofmeister's work was the recognition of " Alternation of Generations " in the vegetable kingdom. The phrase was introduced by Steenstrup in 1845 to describe cases " when animals produce offspring which do not resemble their parents but produce progeny like the original parent." In botany the term was at first applied only to the succession of vegetative and reproductive shoots. In 1856–8 Pringsheim drew attention to the fact that in the fresh-water Algae, Oedogonium, and Coleochaete, the fertilised ovum did not produce a new plant directly but gave rise to a cluster of cells each of which became a new plant, and compared this group of secondary oosperms to the sporocarp in the lowest Bryophyta. These observations, however, did not arouse the interest they might have done had they been published after 1860, when Hofmeister's results were available.

Celakowski, in 1868, was the first to appreciate the true bearing of Hofmeister's discoveries. He recognised two distinct stages in the plant's life-cycle, a sexual and an asexual. In the lower plants, Algae and Fungi, he pointed out that the asexual stage was multiple, one asexual generation succeeding another until at length the cycle was closed by a sexual generation, while in archegoniate plants the sexual and asexual generations succeeded each other with perfect regularity. Further, he drew attention to the fact that the two generations in the Thallophyta were similar to each other in all respects, save in the nature of the reproductive organs, whereas the sexual and asexual stages in the Archegoniatae were profoundly different. Celakowski termed the sexual stage the " protophyte " and the asexual one the "antiphyte," and described the alternation of generations in Thallophyta as "homologous" and that in Archegoniatae as "antithetic." It would appear that he did not regard the cluster of cells formed by the division of the oosperm as having any counterpart in the life-cycle of the Archegoniatae, for he speaks of Algae as having three generations.

In 1868 Sachs published his famous textbook, the first really important general statement on botany as a whole since Schleiden's *Grundzüge* in 1842. Subsequent editions were published in rapid succession, and an English version appeared in 1882, based on the fourth German edition of 1874. In this work Sachs upheld the views of Celakowski, and founded a group of Thallophyta called Carposporeae, which included Ascomycetes and Rhodophyceae, in all of which he recognised an alternation between a sexual generation and a parasitic sporocarp. He defined the word "spore" as a reproductive cell resulting directly or indirectly from an act of fertilisation, with or without the intervention of a vegetative phase—"all other unicellular and nonsexual organs of reproduction we shall not term 'spores' but gonidia or conidia." Sachs added: "If the act of fertilisation does not result in the production of any vegetative structure, or the second generation be altogether suppressed, the fertilised oosperm would then become a spore . . . an equivalent for the whole of the second generation." This is of course a somewhat pedantic subordination of the facts to the exigencies of a terminology. The fusion of an ovum and a sperm results in the formation of an oosperm and an oosperm only. If the terminology be defective that does not alter the fact. In lumping together Algae and Fungi that exhibit this carposporic habit Sachs is reviving the old error of classifying on one "predetermined mark" which had hampered phylogenetic taxonomy since the days of Caesalpino.

In 1874 Farlow, an American botanist, made the

discovery that in certain ferns the sporophyte might arise vegetatively from the gametophyte without the formation of any archegonia ; this phenomenon Farlow termed " apogamy." Pringsheim thereupon experimented on Mosses, and obtained protonemata, or vegetative regions of the gametophyte, as buds from the seta of the sporocarp, and this was later termed " apospory." These discoveries led Pringsheim in 1877 to formulate the view that the two stages—gametophyte and sporophyte—in the Bryophyta, were homologous, seeing that one could give rise to the other by purely vegetative methods, and that sporangia were consequently homologous with archegonia and antheridia. The chief point of difference between Bryophyta and Thallophyta, so far as alternation was concerned, would thus be that in the former the series of recurrent sporophytes present in the latter was reduced to one which had become parasitic on the gametophyte.

Celakowski (as interpreted by Sachs) and Pringsheim were in this way the pioneer leaders of two opposing schools in the interpretation of alternation of generations ; as to the facts of course there was unanimity of opinion. In his reply to Pringsheim in 1878 Celakowski introduced the idea of the sporocarp of the Archegoniatae as a newly interpolated generation between the sexual and asexual generations of the Thallophyta, and this concept of an interpolated phase was fully elaborated at a later date by Bower, though in a somewhat different sense.

As Bower's theory in its original form is the one that has been very generally accepted it will be necessary for you to study it somewhat in detail. It is formulated in a long series of most important monographs in the *Transactions of the Royal Society* and in the *Annals of Botany*, from about 1890 onwards, and restated in a more compact and accessible form in his book, *The Origin of a Land Flora*, published in 1908.

You are already acquainted with the fact that in the

life-histories of the Hepaticae and Mosses it is possible to trace, with the greatest readiness, a progressive sterilisation of the sporocarp, starting with such a form as Riccia and passing through Sphaerocarpus, Targionia, Marchantia, Pellia, Anthoceros, and other genera, until we reach the most highly elaborated sporocarps of the Mosses. While in Riccia the great majority of the cells into which the oosperm divides actually become carpospores or secondary oosperms, in such a type as Pellia most of the products of division are sterile and subserve other duties, such as absorption, conduction, protection, dispersal, etc., and only a comparatively small number are strictly reproductive, although all the units may be regarded as potentially sporogenous. That cells of these sterile regions should under certain conditions produce protonemata, as Pringsheim found, was thus only what might have been expected; the cells in question had recovered their original sporogenous capabilities.

Although this idea of progressive sterilisation in the Bryophytic sporocarpia is the one almost universally accepted, it should be borne in mind that other conceptions of the phenomena are also held. Thus Goebel, in 1910, put forward the view that the line of the Marchantiales forms a reduction series, and that Riccia, instead of being the simplest and earliest member of the group, is in reality a reduced type.

In the monographs to which I have referred Bower examined in great detail the structure and development of the sporogenous organs of all the chief genera of Vascular Cryptogams and of many Phanerogams also, with the view of establishing the occurrence of sterilisation of potentially sporogenous tissue throughout the vegetable kingdom, and on the data he accumulated, together with critical comparisons of the conditions existing among the Bryophyta, he formulated his hypothesis, which perhaps I may best give you in his own words.

" The Archegoniate series," he says, " is undoubtedly of Algal origin, and this their gametophytes amply bear out. . . . They probably sprang from filamentous green aquatic forms, inhabiting shallow fresh water or the higher levels between the marine tide-marks. Certain forms spread to the land where access of water was only an occasional occurrence; in these the sexual process could only be effected at time of rains, of floods, or copious dews, and even this might not take place unless the sexual organs were fully mature; thus less dependence could be placed upon sexuality for propagation, and an alternative method of increase of individuals had to be substituted. This was done by the production of the sporophyte from the zygote; once fertilised, a zygote might in these plants divide up into a number of portions (carpospores), each of which would then serve as a starting point for a new individual, and dry circumstances under which they would be powdery would favour their dispersion. In proportion as these plants spread to higher and drier levels (in accordance with the advantage they gained from escape from competition and more free exposure to light for assimilation) the chance of a frequent recurrence of the circumstances necessary for sexual reproduction would be diminished and the dependence on the carpospores for propagation would increase; consequently the number of spores produced by each sexually-formed sporophyte must be greater, if the race is to survive and be in a position to compete. Any increase in the number of spores entails greater supply of external nourishment during their formation; this, in the phylum of the Bryophytes, is chiefly supplied from the gametophyte, which shows distinct adaptation to subaerial habit, while the means of nutrition on the part of the sporophyte itself are in these plants very limited and the external morphology of it very slight. In other distinct phyla, however, such as the Filicineae, Lycopodineae, and Equisetineae, the sporophyte itself assumes the function of nutrition; a higher morphological differentiation of parts followed, and a more clear distinction between the organs which were to supply the nutriment (stem, leaves, roots) and the parts devoted to the formation of spores (sporangia), this for the first time stamped the sporophyte with a character of independence and permanence."

The general survey carried out by Bower and others brought into the foreground three distinct but related problems: (1) The relationship of the so-called homologous alternation of generations in the Thallophyta (combined, as it is in some forms, with the somewhat confusing phenomena of the appearance of the carposporic stage) with the pronounced antithetic alternation of gametophyte and sporocarp in the Bryophyta; (2) the point of origin of the Pteridophytic line with its independent and highly elaborate sporophyte and greatly reduced gametophyte; (3) the homologies between the Pteridophytic reproductive organs and alternation and the condensed and concealed reproductive structures in the Phanerogams.

(1) In the Thallophyta, you will remember, it was recognised that a sexual stage is, as a rule, succeeded by several asexual generations, and that the cycle is finally closed by a sexual stage once more. It had been noted, however, that plants producing sexual organs might give rise to asexual organs also; indeed, long after Celakowski's time it was shown that the formation of carpogonia, antheridia, and tetragonidia on the same plant, and even in the same branch of such a Red Seaweed as Polysiphonia, was by no means a rare occurrence. In 1896 Klebs published the results of twenty years' experimentation, which tended to show that sexual, asexual, and neutral conditions could be induced in many of the lower Thallophyta, both green and non-green, by changing the composition of the culture media, with or without altering the thermal, photic, or other environmental

conditions under which the plants were cultivated. Over and above this development of sexual organs and of gonidia, however, it had been found that in some Thallophyta another mode of multiplication had made its appearance. In Oedogonium, for instance, the fertilised ovum did not become a new plant directly but subdivided into four motile "zoospores," each of which became in due course a new Oedogonium. It was claimed by some authorities, notably by Scott, in 1896, that "the zoospores formed from the oospore on germination are identical with the so-called zoogonidia formed on the vegetative plant at all stages of its growth," and holding this view, Scott, and others who agreed with him, rejected the theory that suggested that such a subdivision of the oosperm indicated the first beginnings of what was destined to become the Bryophytic sporocarp. The two theories may be very briefly summarised and contrasted by saying that in the antithetic theory of Celakowski and Bower the essential feature is "the recognition of the spore-producing generation as an interpolated phase in the life history . . . and not as derived by the modification of a generation resembling the sexual one," while according to the homologous theory "the spore-producing generation in Bryophyta and Pteridophyta is strictly homologous with the sexual generation and the alternation in these groups differs only in its regularity from the succession of sexual and asexual forms in Thallophyta."

The problem was complicated still further by the discovery of a remarkable cytological difference between the gametophyte and sporophyte. It had been known for some time that nuclei in the course of their division always exhibited a definite number of chromosomes peculiar to the cells of the specific plant under consideration. In 1893 Overton found that the nuclei of the cells of the prothallus of Ceratozamia, one of the Cycadaceae, had only half the number of chromosomes present in the nuclei of the cells of the sporophyte. In 1894 Strasburger expressed the view that this difference in the number of the chromosomes was universal between the gametophyta and sporophyta of all Archegoniatae, so that if the number in the former phase be represented by x then the number in the latter is $2x$. A reduction at some stage in the life cycle from $2x$ to x was thus essential, and Strasburger fixed this halving at the point where the spore mother cell gave rise to the carpospores. Incidentally it may be noted that this reduction division may take place somewhere else, *e.g.* before the formation of ovum and sperm, as in the case of Fucus, where no asexual reproduction nor carpospores occur. Of course, if this cytological distinction could be shown to be universal and fundamental and not merely physiological and adaptive, it would give us a criterion by which the homologies between the two types of life cycle might be settled.

The antagonism between the two schools was accentuated by Lang in 1909 when he put forward what he termed an "ontogenetic theory of alternation." "In organisms without any alternation of generations," he says, "there is only one type of body, and any germ cell, whether sexually or asexually produced, can be regarded as a specific cell with the power of giving rise under the proper conditions to a new individual like the parent." In cases where a definite alternation of generations occurs "we meet twice in the life cycle with a germ cell, *i.e.* a cell capable of developing into a new organism. These two germ cells are the spore and the fertilised egg. The result of the development of these two cells may be closely similar or widely different." As illustrations of the case where the two germ cells produce morphologically identical vegetative bodies he gives Dictyota and Polysiphonia, while the Bryophyta and Vascular Cryptogams present us with two very dissimilar generations, the one derived from the spore, the other from the oosperm. Attention is next drawn to the cytological difference between these two generations, and reference is made to Lloyd Williams's work on Dictyota published in 1898, and to Yamanouchi's paper on *Polysiphonia violacea,* which appeared in 1906, where it was shown that although the sexual and asexual generations are morphologically identical, cytologically they differ profoundly, the sexual generation, produced from a spore, having x chromosomes in its nuclei and the asexual plant produced from the oosperm having $2x$ chromosomes. Lang considers that these Algae "have an alternation of generations strictly comparable to those in archegoniate plants." (The crux of the whole position lies in that sentence—are these generations comparable with the sporophyte and gametophyte of the Mosses and Ferns ?) "In seeking for an explanation of the great differences between the alternating generations which are characteristic of the Bryophyta and Pteridophyta it is evident that there are two possible views : (*a*) that the germ cells are so different that they necessarily give rise to bodies of different structure ; (*b*) that the two germ cells alike represent specific cells of the plant, but that the conditions under which they develop are so different that two very unlike bodies result." Lang accepts the second of these alternatives and goes on to explain his theory in the following words : "There is normally a great difference in the conditions under which the spore and the fertilised egg commence their development in all archegoniate plants. The spore develops free, in direct relation to the soil, water, light, etc. The fertilised egg, on the other hand, develops in relation to the body of the sexual generation. It thus develops, under profoundly different conditions from the spore, firstly in that it is removed from all the influences acting on the spore, and secondly in that it is exposed to a new set of nutritive and correlative influences proceeding from the maternal body. . . . In the case of the spore-producing generation of Bryophyta this removal from external influences and exposure to the maternal influence of the sexual generation last practically throughout the development of the sporogonium. In the Pteridophyta the asexual generation ultimately becomes free from the prothallus. In all cases however the development is initiated and has advanced to the establishment of the various organs of the sporophyte under the maternal influence. We are justified in assuming from some particular cases which have been studied that each stage in the ontogeny is determined by the preceding stage. If, therefore, as in the Pteridophyta, the first steps of development have taken place under the influence of the prothallus, the influence of the preformed parts of the young sporophyte may be legitimately assumed to exercise a 'formative induction' on the further course of its development."

Lang's statement of his theory aroused an important discussion at a special meeting of the Linnean Society in February 1909, which was taken part in by supporters of both the homologous and antithetic hypotheses, but the views expressed, as also the critical paper published some time afterwards by V. H. Blackman, I must leave you to read for yourselves. Farmer, one of the botanists present at the Linnean discussion, while admitting the importance of Lang's work, denied that "any insight was afforded thereby into the causes which, in the first instance, were responsible for the cyclical alternation, or that the rival claims of the homologous and antithetic theories were capable of being decided on such lines."

Bower's present position may be best understood from the following sentences extracted from his recently published book, *Botany of the Living Plant*: "The origin of the sporophyte in the Archegoniatae is quite problematical, since no certain ancestry is known for them. It has been suggested, on the one hand, that the dependence of the sporophyte upon the gametophyte s a true indication of its derivative origin. In that case the sporophyte may have originated from a post-sexual production of spores, and have been originally without vegetative tissue: the latter being produced in the first instance by sterilisation of some of the potential spore mother cells. Comparison among the sporogonia of Liverworts and Mosses makes it appear probable that such *sterilisation* has actually occurred in them. On this theory the simple sporogonium of Riccia, which has no distinction of apex and base, would represent a primitive state, and more elaborate capsules may be seriated progressively in relation to it. A good case can be made out for an alternative view: that the Archegoniatae sprang from some Algal ancestry which already possessed alternate but separate generations, of which the sporophyte developed from a freely shed ovum. That, on adopting a land habit, the ovum, ancestrally fertilised as it lay free in the water, was retained in the parent gametophyte, in a protective organ, the archegonium; and that the *encapsulation* had its effect in moulding and modifying the young sporophyte to such forms as are seen in the simplest Bryophytes. In the present state of actual knowledge such suggestions are little better than speculations."

(2) To settle the precise point of origin of the Pteridophytic line with its independent sporophyte was an even more difficult problem. If we elect to follow Bower's original interpolation theory it is manifest that it is amongst the Hepaticae, if anywhere among existing forms, that we must look for this starting point, and if we consider such a type as Anthoceros it does not seem outside the bounds of imagination to conceive of the sporocarp, already fairly highly organised, as striking root for itself into the soil through the ventral layers of the gametophyte. In the sporocarp of Anthoceros there are all the possibilities of development of the sporophyte, for it possesses a meristem, stomata, and photosynthetic

tissue, the first beginnings of a vascular system and almost the initials of sporangial regions. Beyond the extension of the haustoria into roots, all that is wanted is the enlargement of parts of the axis into photosynthetic areas and the association of definite sporogenous tissue with these extensions. We should thus obtain a structure not unlike an Ophioglossum or a Phylloglossum. Campbell is responsible for the suggestion that the former genus might represent the nearest living descendant of an Anthoceros-like ancestor, while Bower at one time favoured Phylloglossum, and figured the primitive Pteridophyte as a plant with a simple erect reproductive axis bearing sporophylls, each with an axillary sporogenous region, a sporangium or sporangiophore.

It would be most misleading, however, were I to leave you with the impression that the main line of evolution of the Pteridophyta had now been definitely established, and that all that was left for the future to accomplish was the elaboration of details. How far we are from having reached such a consummation I must now endeavour to prove to you. In 1916 there appeared a notable memoir by Dr. T. G. Halle, on certain fossil plants from Röragen in Norway, and in it an account is given of a remarkable type which he named Sporogonites. In the following year Kidston and Lang published a monograph on a fossil plant to which they gave the name of Rhynia, from the Old Red Sandstone of Rhynie, Aberdeenshire. In May 1917 Bower contributed an account of these two discoveries to the *Glasgow Herald*, and I cannot do better than quote from his article: "But the real novelty of Dr. Halle's memoir," he says, "lies in the discovery of a new fossil which he describes as Sporogonites. It consists of a simple stalk, bearing a terminal capsule containing spores. Though relatively large, Dr. Halle regards it as a structure closely agreeing with the sporogonium, or capsule of the Mosses, of which it may be a generalised type. He justly remarks that the occurrence of this form among the very oldest remains of land plants (Lower Devonian) hitherto known is very surprising, for, excepting some cases of doubtful interpretation, moss capsules had not been recorded from any horizon earlier than the Tertiary. This was always a stumbling block to the morphological theorist. But now, in the presence of this proof of the extreme antiquity of the Bryophyte type, he will see new possibilities open before him of the origin of vascular plants from some moss-like ancestry. Dr. Halle's discovery will certainly bring the Bryophytes again into prominence in theories of descent of vascular plants."

Bower then proceeds to sketch the probable appearance of Rhynia, whose external features and microscopic structure have been marvellously well preserved: "Though clearly a land-growing vascular plant, it had no leaves or roots, but was composed of branched cylindrical stems. Some of these ramified underground, and were attached to the peaty soil by numerous root hairs, often grouped on bosses of the outer cortex. Others grew upwards as tapering aerial stems, which also bore bosses on their surface. The lower of these bore root hairs; others appear to have produced lateral branches easily detached, and serving for vegetative propagation. The cylindrical stems consisted of a central core of conducting tissue, covered by a cortex probably green, and invested by a superficial epidermis with stomata, or breathing pores. At the ends of stout stalks the plant bore cylindrical sporangia, filled with spores of a type usual in vascular plants.

" In its leading characters of nutrition and propagation Rhynia is clearly comparable to such types as ferns, club mosses, and horsetails. Its peculiarity lies in the absence of roots and leaves and in the apparently solitary distal sporangia. In organisation it thus appears to take a place intermediate between moss-like and fern-like plants. This becomes more significant when its early occurrence is remembered, together with the fact that in the discovery of Sporogonites by Halle the early existence of moss-like plants is now established." After referring to work by Kidston and Lang at present in progress, Bower concludes: "The new facts have put theoretical morphology again into the melting-pot, and there is every reason to expect that it will emerge strengthened and refined."

(3) The labours of Hofmeister had done much to demonstrate the continuity between the Pteridophyta and the Phanerogams. The first group to be linked up with the Pteridophyta was the Gymnospermae, and the general tendency, at least from 1890 onwards, was to connect this group more closely with the Vascular Cryptogams than with the Angiosperms. Thus in 1891 Belajeff showed that the Gymnosperm pollen grain gave rise to something that could be homologised with the antheridium formed in the microspore of Selaginella, and this homology was greatly strengthened by the discovery of motile sperms in Ginkgo by Hirase in 1895, in Cycas by Ikeno in 1896, and in Zamia by Webber in 1897.

While the stamen and pollen-sacs of the Cycadaceae could be readily correlated with the microsporophylla and microsporangia of the Pteridophyta, there was nothing among living members of that group that appeared comparable to the ovule and its embryo-sac. Coulter and Chamberlain in their work on the *Morphology of Gymnosperms*, first published in 1910, suggest that "the oldest ovule had a single integument entirely free from the nucellus; in testa-formation this integument differentiated into three layers, the outer fleshy, the stony, and the inner fleshy; the ovule was supplied with two sets of vascular strands, the outer traversing the peripheral region of the nucellus; and the beaked tip of the nucellus broke down more or less completely within the firm and resistant epidermis to form a pollen chamber." I must return to this subject presently in discussing the interpretation of the parts of the flower.

I do not propose to weary you with the speculations that occupied the energies of many morphologists from 1860 onwards in the attempt to establish grades of morphological members in the plant. Starting with the recognition of "caulome" and "phyllome" by Naegeli and Schwendener—an echo of the "nature philosophy" of Wolff and Goethe—the morphologists added "thallome" to their vocabulary, to designate an undifferentiated body where stem and leaf could not be recognised, as in Lemna, and "trichome" for sundry outgrowths from a "phyllome" or a "caulome." The sporangium next gave trouble to the philosophically-minded botanist, some regarding it as coming under the category of "trichome," others considering it as a new structure, something *sui generis*.

It was in the interpretation of the flower that the theorising of the morphologists ran riot. The change of view that makes itself specially evident after the appearance of Hofmeister's papers may perhaps be best expressed by saying that the flower had come to be regarded as a specialised branch bearing spore-producing organs, and thus not strictly speaking as a sexual apparatus at all. The development of the appendicular organs were ably worked out by Payer in 1857, in his magnificently illustrated *Organogénie de la fleur*, and the symmetry of the various members by Eichler in 1875 in a valuable treatise entitled *Blüthendiagramme*.

During the later years of the century a controversy arose over the metamorphosis of the floral leaves. Four categories were recognised, viz. sepals, petals, stamens, and carpels, and the problem that exercised the minds of the morphologists was whether foliage leaves became altered successively into these members—that is to say, whether the metamorphosis was progressive, or whether all the appendages of the floral axis were to be regarded as primitively sporogenous and that petals and sepals were sterile sporophylla; in other words, whether the metamorphosis was retrogressive. The school that took Engler for its leader regarded the primitive flower as composed of one or more naked sporophylla, flowers of higher type arising from them by differentiation of these sporophylla into stamens and carpels and by the addition of a perianth. The school that followed Bower, on the other hand, regarded the primitive flower as a strobilus provided with a perianth which had arisen by sterilisation of the outer or lower sporophylla.

The morphological nature of the female cone of such a plant as Pinus was another of the debated questions of the last fifty years of the nineteenth century, and the matter was again given prominence to by Lotsy in his monumental *Stammesgeschichte,* which is still in course of publication. I do not think, however, that there is much to be gained by detailing all the, often fanciful, interpretations that were put forward. Perhaps the explanation advanced by Sachs and Eichler was the one most generally accepted, viz. that the cone is a single flower consisting of an axis with many sporophylla arising from it, each sporophyll being a carpel bearing on its upper surface a large placental scale on which are developed two ovules. It is well, however, that you should bear in mind that Lotsy does not accept an explanation so simple as this. In the volume of his great work entitled *Cormophyta Siphonogamia,* 1911, he says : " Closer examination shews that the female fructifications met with in this group (Coniferae) belong to two distinct morphological types. . . . In Cupressus we have a cone whose axis bears one type of bract or scale only. This scale bears on its upper surface the ovules or megasporangia, and hence is the equivalent of a sporophyll, and the cone is consequently a strobilus or flower. The conditions are quite different in a cone of Abies. Instead of carrying only one kind of scale, the cone axis shews two types of scale, viz. sterile, pointed, narrow scales, the bract scales, and fertile, broad, obtuse scales, the so-called seed scales, which bear the ovules or megasporangia. These seed scales arise in the axils of the bract scales and, since we are unacquainted with any case of a ' folium in axilla folii,' we must conclude that the ' seed scale ' is a modified axillary organ, and hence that the cone of Abies is an inflorescence." Lotsy supports his views with anatomical evidence, but the matter is of too controversial a nature to discuss in a course of lectures like the present.

Under the influence of the views on metamorphosis held by botanists about 1860, the stamen was regarded as a leaf whose half-blades had become transformed into pollen-sacs, but the recognition of the pollen grain as a microspore gradually led botanists to regard the pollen-sac as a microsporangium, a view emphasised by Goebel in 1881 after its anatomy and development had been worked out by Warming in 1873.

The ovule gave far more trouble. According to Schleiden and Braun it was a bud arising from a axial placenta, and consequently the integuments represented fused leaves. In his textbook Sachs gives different morphological significations to the ovule according to its mode of origin and position. A terminal ovule is of an axial nature, a lateral ovule is a modified leaf, a marginal one is a branch of a leaf, while a superficial one is to be included " in the category of such foliar outgrowths as we have already found to occur in the form of sporangia among the Lycopodinae. The ovules of Orchideae must, however, be included under the category of trichomes, in as much as they owe their origin to simple superficial cells of the parietal placentae." By 1881, however, the general opinion was that the ovule was a megasporangium and that the embryo-sac was a spore mother cell, which became a megaspore with or without division, and hence that the cellular tissue ultimately formed in it corresponded to a female gametophyte. This view was put forward by Strasburger, who also held that the gametophyte produced one much reduced archegonium, the so-called "egg apparatus," and that the gametophytic tissue, at first extremely scanty in amount, was stimulated to renewed growth after fertilisation, and formed what we call "endosperm." The synergidae were, on the other hand, interpreted as potential ova by Dodel in 1891, a view that was supported by several investigators who had found that these cells might give rise to accessory embryos.

Fertilisation

During the last thirty years of the nineteenth century the process of fertilisation was cleared up, more especially by Strasburger, who, in 1884, observed the fusion of one of the nuclei from the pollen tube with the nucleus of the ovum, the other pollen nucleus evidently degenerating. In 1898, however, Nawaschin showed that the second generative nucleus did not degenerate, but that it entered the embryo-sac and fused with the so-called "definitive nucleus of the embryo-sac," a phenomenon that was termed by its discoverer "double fertilisation." This gave rise to the view that the endosperm was not a much delayed gametophyte but really a second and greatly modified embryo, an interpretation that was believed to be supported by the peculiar condition known as "xenia." In 1867 Hildebrandt noticed that when yellow maize was crossed with brown maize both the embryo and the endosperm exhibited hybrid characters. It was thought that this peculiarity in the endosperm might be accounted for on the theory that the fusion of the second generative nucleus with the embryo-sac nucleus was sexual in its nature.

To complete the story, Hanstein, Famintzin, and others, between 1869 and 1890, traced the development of the embryo both in Monocotyledons and Dicotyledons, and showed that only part of the oosperm gave rise to the embryo proper, while the remainder acted either as a temporary storehouse of reserve or as a haustorium, while several workers, such as Strasburger, Hegelmaier, and Ganong, brought to light cases of polyembryony, where accessory embryos sprang from the synergidae, from the antipodal cells, or even from the cells of the nucellus. True parthenogenesis, or the formation of an embryo from an unfertilised ovum, was recognised by Juel, in 1898, in Antennaria.

To sum up, you will see that the net result of the labours of a generation of workers was to show that a seed was a structure in which three generations were represented, one sexual and two asexual. The testa, replacing the ovular integuments, was a product of the primary sporophyte, while the nucellus was a solid megasporangium in which only one megaspore was functional. Within the megaspore arose a gametophyte with a much reduced archegonium, from whose ovum, after fertilisation, there developed in part an embryo, the sporophyte of the second generation. Whereas, in the fern, a spore gave rise to an independent gametophyte with both male and female organs, in Selaginella there were two types of spore, one, the microspore, developing a greatly reduced male gametophyte, and the other, the megaspore, forming an almost equally reduced female gametophyte, both of which, as a protection from the plant's great enemy, drought, were endosporal in the development. In by far the greater number of species of Selaginella both microspores and megaspores when ripe escape from their respective sporangia, and fertilisation is effected after dispersal, but in comparatively rare cases the megaspore is fertilised before dispersal. In the Cycad the microspores still escape, but the megaspores are retained permanently in the sporangium, whose wall has taken on a nutritive function, necessitating the development of a new ovular integument. The microspores have thus to be transported to the micropyle, where they find prepared for them a pollen-chamber or cavity filled with fluid into which they shed their sperms. In the higher Gymnosperms and in the Angiosperms the zoidogamic method of fertilisation is abandoned, and in the latter group the megasporangia, or ovules, become further enclosed in the sporophylla, and fertilisation is effected by the generative nuclei, conveyed to their destination by a tube formed by the microspore, which, after the pollen grain has been caught by the stigma, or apex of the sporophyll, has to penetrate the style in order to reach the ovum. The whole problem for the Angiosperm, so far as reproduction is concerned, thus centres round the transport of the pollen grain or microspore from the stamen to the stigma, and the consequent adaptation of the sterile floral leaves to facilitate this transference—a subject that was studied by many workers from the days Conrad Sprengel, through Darwin and Herman Müller, down to a period, only a few years ago, when Knuth gathered all the data together in his great *Handbook of Flower Pollination*, of which an English translation appeared in 1906 and following years.

These are some, but only some, of the first-fruits of the publication of Hofmeister's *Vergleichende Untersuchungen* and of Darwin's *Origin of Species*, and I think you will now agree with me in regarding the year 1859 as a date of as great interest and importance in the history of botany as 1066 is in the history of Britain.

LECTURE VIII

THE TAXONOMY OF CRYPTOGAMS

I HAVE already shown you how, owing to the researches of Naegeli, De Bary, Pringsheim, and others, attention had been more and more attracted to the taxonomy of the Cryptogams, and more especially of the Thallophyta, while the Pteridophyta, due to Hofmeister's work, had quite eclipsed the Phanerogams in importance. Within the last-named group the Candollean system was still accepted as the basis of classification, but attempts now began to be made at formulating schemes expressing the relationship and proportionate status of all groups of the plant world. The first comprehensive presentation of this kind was that put forward by Sachs in his text-book in 1868. He divided plants into five groups, viz. Thallophyta, Characeae, Muscineae, Vascular Cryptogams and Phanerogams, though he admitted that the division was tentative only. He made no attempt at classifying Algae in detail, but, following De Bary, divided Fungi into Phycomycetes, Hypodermi (our Uredineae and Ustilagineae), Ascomycetes, and Basidiomycetes. The Lichens, hitherto regarded as autonomous plants, had by 1868 been shown to be compounds of Algae and Fungi, as a result of the labours of Bornet, Baranetski, and, chiefly, Schwendener. This view of their organisation, though at first almost fiercely opposed by pure lichenologists like Crombie and Nylander, was confirmed by many authors during the next decade, notably by Rees, Stahl, Bonnier, and others.

In the earlier editions of his textbook Sachs laid stress on heterospory in his classification of the Vascular Cryptogams and created three chief divisions, viz. Filices, including all the fern-like plants, save Marattiaceae, which he united with Ophioglossaceae to form a second group, and also Equisetaceae; these were all characterised by being isosporous. To these he added two partially heterosporous divisions, the so-called Rhizocarpeae (Hydropteridae) and Lycopodiaceae. Later on Sachs made several fundamental changes in his scheme, the chief effect of which was to unite the Algae and Fungi into one series, subdivided according to the nature of the reproductive organs. The lowest groups he termed Protophyta, which included all the forms we now call Schizophyta; the second group was the Zygosporeae with isogamous reproduction; the third the Oosporeae with differentiated ova and sperms; and the fourth the Carposporeae, where the fertilised ovum gave rise to a cluster of carpospores. In this last group Sachs placed the Characeae. The Vascular Cryptogams were also reclassified on the ground that heterospory was not a sufficiently fundamental criterion on which to base the larger divisions. The Rhizocarpeae were now grouped with the Filicineae, and the Lycopodiaceae, Selaginellaceae, and Isoetaceae were united under the name Dichotomae, from the supposed method of their branching.

In 1880 Goebel introduced a new idea into the taxonomy of Vascular Cryptogams by emphasising the difference in the mode of origin of the sporangium in the several members of the group. He established Leptosporangiatae for those in which the sporangium sprang from a single cell, and Eusporangiatae for those in which the sporangium sprang from a cluster of cells—another illustration of generalisation on the basis of a single mark. The terms are still in use, but mainly in a descriptive sense.

The next change was put forward by Eichler in 1883,

and his scheme really formed the groundwork on which Engler based his system which has become so widely adopted by continental and British botanists as elaborated in the *Natürliche Pflanzenfamilien*, published by Engler in association at first with Prantl, from 1887 onwards. The chief features of Eichler's scheme are that he rejects Sachs's arrangement of Algae and Fungi in two parallel series and institutes two separate classes, Algae and Fungi, under Thallophyta, uniting that larger division with Bryophyta and Pteridophyta into a sub-kingdom Cryptogamae, as equivalent in rank to the Phanerogamae. The modifications introduced by Engler we will discuss later on when we come to review the state of taxonomy in the present century.

Taxonomy of Phanerogams

So far as Phanerogams are concerned the system established by De Candolle was the one in practically universal use up to 1860, but after that date considerable modifications were introduced by two men whose names are exceedingly prominent in the history of botany, viz. George Bentham and Joseph Dalton Hooker.

George Bentham was born in 1800, and, while still a mere boy, made the acquaintance of such distinguished botanists as Robert Brown, John Lindley, and W. J. Hooker. In 1830 he became friendly with Alphonse de Candolle, and began to collaborate with him in the publication of the famous *Prodromus*. Shortly afterwards he came in touch with Joseph Hooker, with whom he worked during the later years of his long life. In 1854 he was induced to associate himself with the Herbarium at Kew, then under the direction of Sir W. J. Hooker, and began the preparation of the first of a series of floras of the British Colonies, the publication of which had been contemplated for some time by the Kew authorities. In the course of his labours on these floras Bentham found himself constantly in difficulties with the precise limitations of genera, and this led him to undertake his *magnum opus*, the *Genera Plantarum,* a work in which, at a later date, he had the invaluable assistance of Sir Joseph Hooker, the distinguished son of the then Director of Kew, Sir W. J. Hooker.

The fundamental principle on which Bentham and Hooker worked was to start with the genera and smaller associations and allow more general relationships to develop gradually and lead them to higher groupings.

The scheme ultimately adopted was the following :

A. Dicotyledons.

1. Polypetalae.
   Thalamiflorae ; Disciflorae ; Calyciflorae.
2 Gamopetalae.
   Inferae ; Superae ; Dicarpeae.
3. Monochlamydeae or Incompletae.
   Curvembryae ; Multiovulatae Aquaticae ; Multiovulatae Terrestres ; Micrembryae ; Daphnales ; Achlamydosporeae ; Unisexuales ; Ordines Anomali.

B. Monocotyledons.

Microspermae ; Epigynae ; Coronarieae ; Calycineae ; Nudiflorae ; Apocarpae ; Glumaceae.

The special features of this classification are the institution of a group of Disciflorae between De Candolle's Thalamiflorae and Calyciflorae, and a totally new arrangement of the Incompletae, plants with flowers possessing a single perianth or none at all, for the recognition of these as reductions from higher " complete " flowers was not as yet at all a familiar conception in the botanical world. We would not dream nowadays of placing plants, otherwise related, in two different categories simply because one was aquatic and the other terrestrial, nor would we regard unisexuality as a criterion of more than subordinate importance.

As the *Genera Plantarum* appeared at intervals between 1863 and 1880 it must seem somewhat remarkable to you that there is not the slightest expression in the work of the evolutionary ideas that were then influencing so profoundly all other branches of the science, and notably the taxonomy of the Cryptogams, which Bentham and Hooker did not touch. But Bentham was a firm believer in the dogma of the constancy of species, and he explains his position in the following words: "I can scarcely think that due allowance is made for those who like myself through a long course of study of the phenomena of organic life had been led more and more to believe in the immutability of species within certain limits, and have now felt their theories rudely shaken by the new light opened on the field by Mr. Darwin, but who cannot surrender at discretion so long as many important outworks remain contestable." This is Bentham's Apologia in 1863, but ten years afterwards he felt himself driven to accept the new doctrine, too late, however, to make feasible any attempt at modifying the plan of the *Genera*. Probably his fellow worker, Hooker, who was a consistent supporter of Darwin, had no small share in bringing about the conversion of the veteran systematist to the new views.

I must not omit to remind you that it is to Bentham that you owe the *Handbook of the British Flora* whose aid you are accustomed to seek on all your field excursions, a book, by the way, with the writing of which Bentham is said to have "amused himself before breakfast"!

The other attempts at classification that were made during the concluding years of the nineteenth century need not detain us long.

The discovery by Treub in 1891 that in the Australian genus Casuarina the pollen tube enters the ovule by the chalaza and not by the micropyle in the usual way led to the hasty and ill-considered proposal of Eichler to subdivide the Angiosperms into Chalazogamae and Acrogamae, but this idea was soon abandoned after the discovery of the same phenomenon in Betula, Juglans, and other genera by Nawaschin and Benson. Van Tieghem also, in 1898, put forward a fanciful and quite artificial classification, based chiefly on the anatomy of the root and on certain features of the corolla and the state of development of the seed, but this scheme also very soon died a natural death.

The more peculiarly phylogenetic systems that are associated with the names of Engler, Hallier, and Lotsy we will discuss later.

Progress in Palaeophytology

I must now invite your attention to the progress that had been made in a department of botany that you have already learnt had made its first appearance in the earlier years of the nineteenth century, viz. Palaeobotany. I have already referred to the works of Brongniart, Goeppert, and others, and I may also remind you that Robert Brown published a paper on a fossil Lycopod about the same period. Another eminent palaeobotanist was Binney, who described the fossil trees Sigillaria and Calamodendron, which, on the ground that they presented secondary thickening, he regarded as Gymnosperms, a view that was shared by Brongniart. Carruthers also studied the fossil Equiseta and Lepidodendra in 1868, and founded the genus Bennettites to include certain stems and fructifications which he imagined were of Cycadean affinity, and which were destined many years afterwards to play an important part in the controversy concerning the origin of the Angiosperms. Twenty years later Solms Laubach discovered in these fructifications a dicotyledonous embryo enclosed in an exalbuminous seed. The whole anatomy of this important Mesozoic plant and its affinities with the Angiosperms was worked out by the American palaeobotanist Wieland in the beginning of the twentieth century, but his conclusions may be left for future discussion.

The names that stand out most prominently in fossil botany between 1860 and 1890, however, are those of Williamson in England and Renault in France. Williamson's name is inseparably associated with the plants of the coal measures, and to him and to his coadjutor of later years, Scott, we owe a long series of monographs that rank in importance with those of Hofmeister and Bower as classics in the history of our knowledge of the Vascular Cryptogams. Williamson devoted himself more especially to proving, ultimately with complete success and in opposition to the French school, that the stems of Calamites, Lepidodendron, Sigillaria, etc., although they exhibited secondary thickening, were neither Angiosperms nor Gymnosperms but genuine Vascular Cryptogams. He further showed that the fossils known as Stigmaria were not only roots of Sigillaria but also of Lepidodendron. He also worked out the anatomy of a fossil stem called Lyginodendron, which had been discovered by Binney in 1866, and emphasised its Pteridophytic relationship, while admitting at the same time that it showed distinct Cycadaceous affinities. This type was destined some years later to form the basis of an entirely new group of plants, the Cycadofilices or Pteridospermae, but the full story of the recognition of this family we shall discuss later on.

In France Renault's name is associated first of all with the anatomy of the fossil Pteridophyte known as Sphenophyllum, a genus found later to be allied to the Equisetales, and also with Sigillaria, the structure of which he investigated with Grand 'Eury in 1874. To Renault also we owe much of our knowledge of the most primitive ferns, the Botryopteridae, and of many fern-like fronds that were regarded at that date as allied to Marattiaceae but that were afterwards shown to be the foliage of Pteridosperms. Further, also in company with Grand 'Eury, Renault laid the foundations of our knowledge of the Cordaitales, a group of plants that showed characters allying them with both Cycads and Conifers.

Although undoubtedly interest in the realm of fossil botany during the period of which I am speaking centres in the discoveries made in the rich strata of the older Palaeozoic rocks known to geologists as the Devonian and Carboniferous, the younger strata were also explored. One of the most important finds from these latter rocks was the type of fern frond named Glossopteris by Brongniart in 1828. Further investigations showed that the flora of which Glossopteris was the type was very widely spread over the southern hemisphere during the epoch when the coal measures were being laid down in the north, and, before the close of the Permian period, had spread northwards, overlapping the forest vegetation composed of Lepidodendra, Calamarieae, and such-like giant representatives of our humble clubmosses and horsetails. The study of the distribution of this Glossopteris flora opened up points of profound interest both climatic and geographical, but into these questions I have no time to enter.

Palaeobotany had now become so important a branch of the main science that it demanded a textbook to itself, and so in the later years of the nineteenth century a general compendium of the subject appeared from the pen of Solms Laubach of Strasburg, of which an English translation was published in 1891 by the Oxford Press. Although valuable as forming a work of reference and a summary of what had been discovered up to that date, and as showing how important the study of fossil plants was to the morphologist and taxonomist, it was unfortunate that the work, so far as its mode of presentation was concerned, was so lacking in vitality and vividness in expression, and in this respect stands in marked contrast to such a book as Scott's *Studies in Fossil Botany*, which is so familiar to you all.

You will now, I hope, be able to appreciate how urgent was the need for an entire revision of the taxonomy of plants, so that these extremely important new types of vegetation, even though long extinct, should find a home in the scheme. For instance, it had been almost an article of faith among botanists that the Palaeozoic was the age of ferns, but the discovery of the Cycadofilices revealed the possibility that these fern-like fossils were of far higher rank than was at first thought, and that the bulk of our living ferns, so far from being primitive, were really of comparatively recent origin. The problem of the evolution of the seed habit in Spermatophyta was now in a fair way to be solved. The gaps between the various branches of the Vascular Cryptogamic phylum began to be bridged over, and the affinities of the living types received new interpretations. If any further evidence were required to overthrow the old dogma of the constancy of species, and to support the newer views that the plant world was itself a great tree, with its root deep sunk in the immemorial past, with mighty branches now completely extinct, or forerunners of delicate twigs persistent to the present day, then the witnesses in the rocks proffered this evidence in abundance. Yet even now we have only, as it were, scratched the uppermost layers of some of the strata that form the earth's crust; what may not deeper and more extensive excavations reveal?

Progress in Cryptogamic Morphology

As you have already seen, the second half of the nineteenth century was particularly fruitful in research among the Pteridophyta, but in addition to the more comprehensive works of Bower on the *Morphology of Spore-producing Members*, of Campbell on *Mosses and Ferns*, and of Goebel on the *Comparative Development of*

Sporangia, quite a library of monographs appeared on the anatomy of individual genera or on special organs. Even to mention the contents of these in the briefest possible manner would occupy time out of all proportion to what I may legitimately allow myself, and at the same time maintain some sort of balance among the variety of subjects I have to bring to your notice. I must, however, mention a few of these works, if ever so briefly. There are, for instance, the researches on the anatomy of the Ophioglossaceae by Prantl, Farmer and Freeman, Russow, Jeffrey, and Campbell; on various genera of the Marattiaceae by Farmer, Campbell, and Luerssen; on the Osmundaceae by Prantl and Kny, and on the Gleicheniaceae and Schizaeaceae and on the Hymenophyllaceae by Poirault, Prantl, and Goebel. The group long known as Rhizocarpeae, from the supposed origin of the sporocarps from their roots, was completely investigated not only by Hofmeister and Pringsheim but also by Prantl, Russow, Hanstein, Strasburger, Belajeff, Campbell, and others, with the result that these peculiar aquatic Pteridophytes were shown to be ferns, and renamed Hydropteridae, while the pronounced heterospory exhibited by them turned out to be a case of parallel development in the fern series analogous to, but in no way connected with, the corresponding phenomenon in the Lycopodinae.

In the Lycopodinae also much important work was accomplished on the prothallus and gametes by Millardet, Mettenius, Belajeff, Bertrand, Pfeffer, and Campbell, but especially by Treub and Bruchmann, while the vegetative organs of the sporophyte were investigated by Strasburger, Solms Laubach, Goebel, and others. Isoetes was dealt with in great detail by Russow, and later by Farmer and Campbell, while the Equisetaceae had many points in their anatomy and development cleared up by the researches of Pfitzer, Famintzin, Van Tieghem, Cormack, and Jeffrey. Similarly the anatomy of the genus Selaginella was worked out to some extent by Russow and De Bary, and special points received attention from Janczewski, Treub, Haberlandt, Dangeard, and Le Clerc du Sablon. If the personal reference be pardoned, the comparative anatomy of a large number of species was investigated by myself between 1893 and 1902.

A mere catalogue of names such as I have just given you can have but little meaning unless you take the trouble to study the individual monographs for yourselves, or at least the summaries of their contents from such references and quotations as you will find in Campbell's *Mosses and Ferns* or in Bower's *Origin of a Land Flora*, the best book I know of on the subject, and which I earnestly recommend you to study with the utmost care. If I were to omit all notice of the authors whose names I have quoted you might think I regarded them as of little importance, or blame me for overlooking them altogether when you encountered their names in other works. To mention them without discussing the results they achieved is, I frankly admit, a compromise, but one demanded by the exigencies of time and space.

The Mosses and Liverworts did not attract nearly so much attention during the period of which I am speaking. The outstanding name is undoubtedly that of Leitgeb, who published several important researches on the Hepaticae between 1868 and 1886. To Farmer we owe also much of our knowledge of the cytology of the groups, especially the discovery of centrospheres associated with the nucleus during division, a subject that had attracted the attention of the zoologists in relation to cell division in the animal kingdom. Pringsheim and Stahl also studied the protonemata of the Musci and the mode of origin of the gametophores from them, while Kienitz-Gerloff elucidated the structure of the sporogonium.

Very considerable progress was also made in our knowledge of Algae, both systematically and anatomically. Thus, in addition to the *Species, Genera et Ordines Algarum* of J. G. Agardh, and Kützing's *Tabulae Phycologicae*, we have such important general works as Harvey's *Phycologia Britannica* and *Phycologia Australis*, Reinke's *Atlas deutscher Meeres-Algen*, De Toni's *Sylloge Algarum*, Kjellmann's *Algae of the Arctic Sea*, and Hauck's *Meeres-Algen* in Rabenhorst's great compendium *Kryptogamen-flora*. The morphology and life-history of specific groups of genera were also dealt with in detail in Areschoug's *Observationes Phycologicae*, Agardh's *Till Algernes Systematik*, and Bornet and Thuret's magnificent volumes, the *Notes algologiques* and *Études phycologiques*. Among the more special monographs I must at least mention Berthold's researches on the sexual reproduction of the Phaeophyceae, published in 1881, Schmitz's investigation into the fertilisation of the Florideae, 1883, and Sirodot's paper on Batrachospermum, 1884.

In the Fungi De Bary's fundamental treatise on the *Morphology and Classification of Fungi, Mycetozoa and Bacteria* appeared in 1881, and in it he traced the descent of the Ascomycetes and Basidiomycetes from the oogamous series of the Phycomycetes, these higher groups showing progressive loss of sexuality. Brefeld, however, took an entirely different view of the relationships of these groups, deriving all the higher Fungi from the Mucorinae, a branch of the Zygomycetes, and holding that the fruit of the Ascomycetes was entirely asexual in origin—in short, a cluster of Mucor gonidangia enclosed in a sterile pericarp. Brefeld's views were elaborated in a long series of important monographs, and were placed before the botanical public in a masterly sketch, *Vergleichende Morphologie der Pilzen*, in 1892, from the pen of Von Tavel, who, in this respect, acted as Boswell to Brefeld's Johnson, and his work did much to gain supporters for Brefeld's views. The correctness of De Bary's position was, however, completely substantiated by Harper's researches on Sphaerotheca in 1895. Harper had come to work with Strasburger in the previous year, and had been commissioned by him to make an investigation into the problem of the sexuality or non-sexuality of the Ascomycetes. Strasburger, it should be remembered, was at the time more or less a convert to Brefeld's views. Harper, who, like a certain Israelitish prophet, had been brought to "curse the house of" De Bary ended in "blessing it altogether," for he was able to prove that some of the simpler Ascomycetes at all events were certainly possessed of sexuality, and of course if any single genuine Ascomycete could be shown to possess sexuality Brefeld's phylogeny at once fell to the ground. Many British botanists, however, remained staunch supporters of De Bary's views, and the publication of Harper's work fully justified their conservatism.

The period 1870–90 saw also the complete establishment of the symbiotic theory of the structure of the lichen thallus, and of the correctness of Schwendener's views on the subject. As in the case of the Algae, much special research was carried out on individual genera and on specific questions concerned with the fertilisation of the Saprolegniae and other Phycomycetes, as also on the Schizophyta and the animal-like Mycetozoa, but into the discussion of these researches I cannot enter.

Progress in Cytology

You will remember that in the years between 1850 and 1860 the whole outlook in vegetable, and indeed also in animal, anatomy and histology underwent a complete change owing to the establishment of the cell theory by Schwann and Schleiden, and owing to the identification of protoplasm as the essential constituent of the cell. The years that followed were rich in research on the new structural and physiological problems opened up by these two great generalisations. The investigations were concerned chiefly with (1) the minute structure and chemical composition of protoplasm; (2) the structure of the nucleus and its behaviour during cell division; (3) the structure and chemistry of the cell wall; and (4) the protoplasmic inter-relationship of cells.

Following on the generalisation we now know as the "cell theory," a considerable amount of work was carried out on the differentiation of permanent tissues from primary and secondary meristem on the lines laid down by Hanstein in 1868, and with these investigations we must associate the names of Sachs, De Bary, Van Tieghem, Jeffrey, and Haberlandt. The mechanical tissues and their arrangement in the different organs of the plant were worked out by Schwendener in 1874, while numerous papers were also published on points of detail or on the anatomy of special plant types. Let us look at some of this work a little more closely, beginning with the problems concerned with the cell.

Protoplasm was described by Max Schultze, Von Mohl, and others as a colourless, viscid, granular substance lining the wall, or more or less filling the cavity, of the cell, and exhibiting streaming movements or cyclosis when free to move as in the case of the Mycetozoa. By 1875 it was generally accepted that protoplasm consisted of a framework or reticulum of delicate fibrillae, whose meshes were filled with a fluid "enchylema," the nodes on the reticulum giving the appearance of granulation. The most important modification of this view was that advanced by Bütschli in 1889, and known as the "foam theory." According to Bütschli the protoplasm is to be likened to a froth made by rubbing up oil with a solution of potash or cane sugar, which, under the microscope, presents the appearance of a network, the threads being the optical sections of the walls of the droplets. In this foam numerous granules are suspended, especially at the angles where the droplets touch each other. The free surfaces of protoplasm were observed to be of a denser consistence than the remainder, and were distinguished by Hanstein as composed of "ectoplasm" in contrast with the inner more granular "endoplasm." Pfeffer extended this idea by stating that all exposed protoplasmic surfaces were covered by "plasmatic membranes" of different composition from the rest of the living substance.

The contents of the different segments of a sieve tube had been shown by Hanstein in 1864 to be in continuity with each other through perforations in the sieve plate, and this had been confirmed by Russow and Strasburger in 1882, while protoplasmic continuity between the cells of certain Algae had also been demonstrated by several investigators. In 1897 Gardiner, by the use of improved methods, was able to show that protoplasmic intercommunication was a characteristic of all living cells.

By 1880 the universal presence of a nucleus, discovered by Robert Brown fifty years before, had been accepted as a fact, and its importance in cell-division had come to be recognised, but its structure and functions were as yet very imperfectly known. Strasburger and Flemming were the first to study it in detail and to show that it consisted of a homogeneous or finely granular matrix in which lay imbedded a reticulum or greatly coiled fibre. One granule much larger than the rest received the name of nucleolus. The reticulum itself consisted of a non-stainable basis called achromatin or linin, with imbedded granules of chromatin. The nucleus as a whole was also seen to be enclosed within a nuclear membrane. Chemically the nucleus was composed essentially of a compound rich in phosphorus known as nuclein, and this again was analysable into albumin and nucleinic acid. Flemming and Van Beneden had demonstrated the occurrence of an accessory body, the centrosome, in animal cells, but, in spite of Guignard's attempts to prove its existence in Phanerogams, plant nuclei, save those of Hepaticae, in which Farmer proved their presence, and, according to Williams, Mottier and others, those of Algae, appeared not to possess such bodies.

At the same time great progress was made in unravelling the mysterious changes that occurred in the nucleus during cell-division, changes which were spoken of as karyokinesis or mitosis. Here again we owe the foundation of our knowledge to the labours of Flemming, Van Beneden, and Strasburger. Reduction division was first discovered by Van Beneden in the origin of the sexual cells of the worm Ascaris, and later by Overton in a Cycad, while, as I have already told you, Strasburger in 1894 formulated the generalisation that while the number of chromosomes into which the reticulum subdivided was constant for each plant type, that number was halved in the gametophyte stage, to be reassumed on the fusion of the gametes and maintained throughout the sporophyte, until, in the mother spore cell, reduction division introduced the gametophyte number once more.

Another subject that was hotly debated during the later years of the nineteenth century was the structure and mode of growth of the cell wall. Naegeli, in his classic work on the starch grain, published in 1860, had put forward the view that organised bodies consisted of "micellae" or invisible crystalline particles, separated from each other by equally invisible shells of water, and that the density of the body as a whole was determined by the thickness of these aqueous shells. Growth in thickness, according to Naegeli, depended on the interpolation of new micellae between those already present, by a process which he termed "intussusception." Later investigators, such as Strasburger, held on the other hand that the particles were not crystalline, and that they were bound together chemically and not by molecular attraction, as was suggested by Naegeli, while growth in thickness was effected by apposition of new layers, not by intercalation of new particles.

In 1896 Wiesner advanced yet another theory of the structure of the cell wall. He suggested that the wall was living, being composed of "dermatosomes," as he termed them, connected by protoplasmic threads, and that growth resulted from the formation of new dermatosomes from the included protoplasm. In the last years of the century the intussusception theory had regained much of the support it had lost, more especially when Strasburger, in 1891, to a large extent abandoned his previous views on apposition.

The chemistry of the wall was carefully studied by Mangin in 1892. Previous to that date it was generally held that the wall, at least in its juvenile condition, was composed of cellulose chiefly, and that the middle lamella between any two cells was formed of pectic compounds. It was also recognised that the cellulose became altered by the deposition in it of suberin, cutin, or lignin, or by its transformation into these bodies. Mangin's work dealt chiefly with the pectic constituents, some of which he showed were neutral in reaction, such as pectose and pectin, while others were acids, such as pectic and metapectic acids. In cell division, according to Mangin, the first appearance of the new cell wall is a film of pectates, on both sides of which layers of cellulose and pectose compounds are deposited. De Bary, in 1864, pointed out that the walls of fungal hyphae did not give the well-known blue reaction with iodine and sulphuric acid so characteristic of cellulose, and Wisselingh, in 1898, showed that in Fungi the cell walls are composed chiefly of chitin.

The Origin and Development of Tissues

After the establishment of the cell theory attempts were made to follow the various changes that took place in the differentiation of the primitive homogeneous cells of the embryo and of the growing points into permanent tissues. I have already drawn your attention to the work accomplished by Naegeli in 1858, and by Sanio in 1863, in this relation, but it is to Hanstein that we owe, in 1868, the first clear account of this subject. He claimed to have demonstrated in the growing regions of the Phanerogamic stem and root three definite histogenetic layers of cells, comparable with the germinal layers of the animal embryo, and to these he gave the names of dermatogen, periblem, and plerome, giving rise respectively to the epidermis, the cortex, and the vascular cylinder. To these, in 1874, Janczewski added a fourth, the calyptrogen, as the initial layer of the root cap.

A more detailed examination of different plant types, however, led to the discovery of various modifications of Hanstein's original scheme, some of the germinal layers being shown to have a common origin. In 1877 De Bary in his *Comparative Anatomy of Phanerogams and Ferns* reviewed the whole subject, and came to conclusions somewhat adverse to those of Hanstein and Janczewski. His criticism was as follows: "To the question whether in all cases only definite zones of meristem give rise to definite sorts of tissues, the most general answer according to our present knowledge is a distinct negative. To be sure this negative does not hold for all single cases. For instance, for the large majority of roots, not only does each of the layers of meristem correspond to a definite section of a definite system of tissue, but even the separate parts of each of these sections may often be traced back to its separate initial cells in the apical meristem. . . . But even in roots exceptions occur. The epidermis, for instance in the Gymnosperms, does not originate from a distinct dermatogen layer. . . . The negative of the constant genesis of different sorts of systems of tissues from definite zones of primary meristem holds to a much greater extent in leaf-forming shoots. Here also it is true there are such relations. The system of vascular bundles of many stems of Phanerogams, for instance, is derived exclusively from the plerome cylinder. But exactly the opposite also occurs . . . and the whole of the tissues and tissue systems of the leaves, which are continuous with the similar and synonymous tissues of the plerome of the stem, are formed, according to the data in hand, outside the plerome, being derived, as is the whole leaf, from the periblem and dermatogen, or from the layers of meristem corresponding in position to these."

In the Bryophyta and Pteridophyta the growing points were known to be occupied by a meristem which, in the great majority of cases, could be traced to the regular partition of a single apical cell, and attempts were made to homologise the conditions existing in these plants with Hanstein's layers as reported in Angiosperms. Scott, in 1894, sums up for us the general conclusions arrived at in the following words: "We may safely conclude that, with the solitary exception of the origin of the root cap, the first tangential cell walls do not constantly mark out the main histogenetic layers from which the principal systems of tissues are derived. Such layers arise in different ways in different cases, and at various distances from the apex."

Seeing that Hanstein's classification of tissues on an ontogenetic basis had not proved entirely satisfactory, Sachs next formulated a physiological scheme, distinguishing epidermal, fibrovascular, and fundamental systems derived from a uniform apical meristem, but he also was adversely criticised by De Bary, who said "it did not answer its purpose, which is to serve as a basis for a uniform exposition of the various differentiations of plant tissues."

The Stele Theory

The year 1870 saw the publication of an important work by the French botanist, Van Tieghem, which laid the foundation of an entirely new way of looking at the architecture of the plant axis, and one which, though in a modified form, holds the field to-day. This is what you know as the "stele theory" Van Tieghem began his researches by giving a comparative account of the anatomy of the root and showed that De Bary's "radial bundle" was in reality a cylinder composed of several bundles, the strands of phloem alternating with the strands of xylem, which latter might meet in the centre or leave a more or less lignified pith. These strands were connected with each other by conjunctive tissue, and the whole was enclosed by a pericycle abutting on the outside on an endodermis which Van Tieghem regarded as the innermost layer of the cortex. To this central core, from the pericycle inwards, he gave the name of "stele," a column. In succeeding papers Van Tieghem applied the idea to the stem and claimed that there also a "stele" could be recognised, but that the conjunctive tissue was subdivided into medullary rays and pericycle, and that these tissues were quite distinct in origin from the cortex. At the same time he pointed out that the outline of the "stele" in the stem was not so clearly marked, owing to the disturbance caused by entry of leaf traces and to other causes.

Van Tieghem, who had associated with himself one of his pupils, Douliot, next proceeded to work out all the different conditions he could find among plants and attempted to fit them into his scheme. When the axis possessed a large medulla he found that the endodermis and pericycle might push their way in between the bundles and isolate them from each other, the "monostelic" condition thus becoming "astelic," and the conjunctive tissue of the medulla becoming confluent with and indistinguishable from the cortex. Should these isolated cords again fuse the condition known as "gamodesmy" resulted. In Ferns and in a few Angiosperms, the "stele" in the hypocotyledonary axis divided into two or more branches and for this condition Van Tieghem coined the term "polystely" or "dialystely." Finally, the fragments of the "stele" that entered the leaf he described as "meristeles." In 1891

Van Tieghem received strong support from Strasburger, who, however, suggested the term "schizostele" to indicate the branches that enter the petioles, and also the segments of the "monostele" in Van Tieghem's "astelic" condition.

Some years later the American botanist, Jeffrey, brought forward various criticisms of Van Tieghem's views and also formulated a modification of his own. In his *Morphology of the Central Cylinder in the Angiosperms*, published in 1900, he writes: "The polystelic type of Van Tieghem is not characterised by the repeated bifurcation of the epicotyledonary stele, but there is primitively in the young stem of this type a tubular concentric stele with foliar gaps subtending the points of exit of the leaf traces. The astelic type of Van Tieghem does not result from the separation of the constituent epicotyledonary stele into its constituent bundles, for in the young so-called astelic axis there are no bundles present at all, but a collateral stelar tube with foliar gaps subtending the leaf traces, through which the internal and external phloeotermal sheaths communicate. The medullated monostelic type of Van Tieghem does not originate, as he states, by dilatation of the epicotyledonary stele and the formation of an intrastelar pith, for in favourable cases the so-called medullated monostelic central cylinder of the older stem may be seen to be derived from the so-called astelic condition of the young axis by the degeneration of the internal phloeoterma. Van Tieghem's three types of central cylinder indicated above are all modifications of a single type, which has been designated by the writer 'siphonostelic.' In this type the central cylinder is primitively a fibrovascular tube with foliar lacunae opposite the point of exit of the leaf traces. In the so-called polystelic modification the central cylinder has internal as well as external phloem; in the astelic type of axis the internal phloem is absent. The medullated monostelic type of Van

Tieghem is derived from the last named by the degeneration of the internal phloeoterma, or endodermis. The siphonostelic type of central cylinder as defined above is probably to be regarded as the result of the mechanical strengthening of the cauline axis to enable it to support the palingenetically large leaves which are characteristic of the Angiospermae, Gymnospermae, and Filicales. In these three groups the siphonostelic fibrovascular cylinder is invariably distinguished by the presence of gaps corresponding to the points of exit of the leaf traces, and in this feature offers a marked contrast to the tubular central cylinder of the Lycopodiales and Equisetales, in which there are no foliar lacunae, but, on the contrary, gaps subtending the branches."

All vascular plants thus possess a central cylinder which is either protostelic or siphonostelic. The protostele has no medulla, but consists of a central xylem surrounded by phloem and pericycle; the siphonostele differs in having a central medulla. On this cylinder are inserted the vascular systems of the leaves and branches which may create gaps in the cylinder wall. If the leaves be small, as in the Lycopodiales, the insertion of their vascular systems causes little or no disturbance, if large, as in the Filicales, prominent gaps are produced which, if the leaves be closely set, may overlap so as to obliterate more or less the tubular character of the vascular axis. Similarly, where the vascular systems of the lateral branches join that of the main axis, ramal gaps are formed, disturbing the continuity of the primary cylinder. To these two conditions Jeffrey applied the terms "phyllosiphonic" and "cladosiphonic" respectively.

The various interpretations of the anatomical structure exhibited by different types of vascular plants were summarised and criticised in 1902 by Schoute in his monograph, *Die Stelär-Theorie*, and also by Tansley in his *Lectures on the Evolution of the Filicinean Vascular System*, published in 1907–8. To these works I must refer you for further details, as also to the numerous papers on the anatomy of Pteridophyta from the pen of Gwynne-Vaughan, by whose recent death British Botany suffered so grave a loss.

In 1884 Haberlandt returned to Sachs's point of view of the classification of tissues in his *Physiologische Pflanzenanatomie*, which, after passing through four editions in the original, made its appearance in English dress in 1914. The idea underlying this work is that the whole architecture of the plant is an adaptation in response to physiological needs, and is thus to be regarded from an entirely different standpoint from that assumed by the pure morphologist. Tissues are classified according to their functions into dermal, mechanical, absorptive, photosynthetic, conductive, and so on. Haberlandt did not aim at upsetting the purely morphological arrangement however, for he recognised that tissues of entirely distinct origin may be modified in the same way to fulfil the same function. The treatise is thus more valuable to the physiologist than to the anatomist, and is especially so to workers in ecological botany, a subject which had begun to be widely studied at the end of the nineteenth century, and which, owing to the labours of Warming and A. F. W. Schimper, assumed such prominence in the early years of the twentieth century.

Just as in the case of the Cryptogams, an immense number of papers were published from 1860 onwards upon special points in tissue structure and arrangement, or on the minute anatomy of individual genera or species, but it would take us far too long even to enumerate these researches let alone to discuss them.

Advances in Physiology

Let us now consider what progress was being made in physiology during this period. As we have already seen, the whole outlook had undergone a profound change after the recognition of the fundamental importance of protoplasm. Seeing that this "physical basis of life" was common to animal and plant it became necessary to compare closely the methods of nutrition in the two kingdoms and to follow out their differences and agreements. Slowly the conclusion was arrived at that the plant, as it were, started its nutritive processes at a lower level than the animal, for it appeared not only to carry out the various nutritive phenomena exhibited by the animal in digesting and assimilating "food," but actually manufactured for itself the very "food" so digested and assimilated, out of raw materials obtained from the soil and the air. The term "food" thus came to have a different meaning according as it was used in connection with one type of organism or with the other, and the want of appreciation of this fact introduced much confusion in the numerous students' textbooks that now began to make their appearance.

Further, it was gradually recognised that plant protoplasm, like animal protoplasm, was sensitive and responded to stimuli in various ways, some identical with those exhibited by the animal, more often in ways totally different, and that the nature of these responses was determined chiefly by two factors, the one the special capacity possessed by the plant of manufacturing "food" from inorganic materials, the other its fixed habit and diffused skeleton.

The number of papers published between 1860 and 1900 concerned with the physiology of plant nutrition was enormous. On the problem of photosynthesis and the composition and functions of chlorophyll alone more than seven hundred researches were published during these forty years; so that one is safe in estimating the output for the whole of plant physiology at several thousands. Very many of these were, of course, mere notes, criticisms, or summaries, but, even if we halve the total body of published work, the amount that must be reckoned with reaches the dimensions of a small library. In these circumstances it is not possible to do more than single out the principal lines of research and the outstanding discoveries in each of them.

Absorption of Water and Salts

Our modern views of the mode of entry of water and salts in solution into root hairs date from the publication of Thomas Graham's classic researches on colloids and crystalloids in 1862. A few years afterwards Traube applied Graham's discoveries to the elucidation of the phenomena of endosmosis in plant cells and carried out a series of experiments with the aid of the " artificial cell" that is associated with his name. This, as you know, is formed by placing a drop of a solution of gelatine in a solution of tannic acid, when a colloidal film of gelatine tannate is produced where the liquids meet. The gelatine solution then attracts water through the tannate film, diluting the gelatine within and causing the vesicle to swell and finally burst, exposing a fresh surface of gelatine to the tannic acid, when immediately a new membrane is formed at the point of rupture. Traube found that the pellicles so formed were permeable to some solutes and not to others, or, in other words, were semi-permeable.

In 1877 Pfeffer greatly extended our knowledge of osmotic phenomena by the use of porous earthenware pots in the walls of which semi-permeable membranes had been deposited. He compared this apparatus with the vegetable cell, and agreed with Traube in believing that the passage of solutions through the protoplasm and the cell wall depended on the relative sizes of the molecules of the salts used and the spaces between the micellae of the membrane. He also estimated the pressure set up in such cells by the entry of various solutions and showed how this pressure maintained turgidity in parenchyma.

In the same year De Vries estimated the osmotic

values of the different cell contents and proved a relationship between osmotic pressure and molecular weight. To De Vries we owe much of our knowledge of plasmolysis, a term introduced to indicate the separation of the protoplasm from the cell wall, when the protoplasmic membrane refused to allow the solute outside the cell to enter the central vacuole, even after it had penetrated the cell wall. In 1884 De Vries showed that the osmotic effect depended not on the dissolved weight of the substance but on the number of molecules dissolved, and that when substances are dissolved in quantities equivalent to their molecular weights they all produce the same osmotic effect. A further discovery was that although this was true of non-metallic compounds it was not true of solutions of metallic salts when these were very dilute, for in these cases the salt underwent partial dissociation and the free ions exercised the same osmotic effect as the entire molecule. This is a matter of considerable importance when we take into account the extremely dilute solutions of minerals in the soil that are presented to the root hairs.

Sachs in 1865 pointed out that the osmotic transference of solutes from the soil to the root hairs and from them to the cortical cells not only induced turgidity in these cells but established an internal pressure in the root which forced the water and its dissolved salts into the wood vessels, and so created "root pressure," and accounted for the phenomenon of "bleeding" from wounds. Several investigators, notably Baranetski, Detmer, and Wieler, were also able to demonstrate that root pressure exhibited both a daily and a seasonal periodicity.

Transpiration

Hales, nearly 150 years before, had noticed that water vapour was given off from a leaf surface much less freely than it was from an equal surface of water, but nothing of importance had been done on the subject from his day until Sachs, in 1860, drew attention to the part played by the intercellular space system in the process He was the first to show conclusively that transpiration was a vital phenomenon, what might be described as evaporation under protoplasmic control. He further showed that the function was closely connected with absorption by the root and more especially with the nature of the salts absorbed. Sachs's conclusions were confirmed and extended in 1876 by Burgerstein, and in 1880 by Vesque. About the same time Haberlandt and Von Höhnel carried out their often-quoted experiments on the amount of water transpired from plants of different ecological types.

The study of transpiration and the variations observed in the amount of water transpired naturally led to investigations of the agents chiefly concerned, viz. the stomata. To Von Mohl is due the credit of having been the first to observe that when the guard cells are turgid the stoma is open, and when flaccid, closed, but beyond suggesting that the variations in turgidity might be due to the pressure of varying quantities of osmotic substances in the guard cells, Von Mohl did not carry the matter further. It was not indeed until physiology had clearly grasped the idea of the plant as a sensitive organism responding to stimuli from without that the opening and closing of the stomata came to be looked upon as an instance in point. The mechanism of the guard cells was first of all examined with great care by Schwendener in 1881, and the fundamental importance of the mode of distribution of the thickening on the walls clearly demonstrated. In lecturing to you on transpiration I warned you to note carefully how the whole efficiency of the stoma as a regulator of the escape of water vapour depends on the structure of the walls of the guard cells, so that I need not recapitulate Schwendener's observations.

In 1886 Leitgeb put forward the view that the periodic

opening and closing of stomata was passive, and that the real agents concerned were the surrounding epidermal cells, whose turgidity and flaccidity exerted an alternate squeezing and pulling effect on the guard cells. These two, to some extent, antagonistic views were at last brought into harmony by Francis Darwin, who, in 1898, showed that both factors were operative, although in his opinion the varying turgidity of the guard cells was the chief agent. Perhaps the most important feature in Darwin's paper was the emphasis he laid on the stoma as a piece of vital machinery, extremely sensitive to all changes in the environment. Once again the explanation was seen to be deducible from the responses given by the protoplasm to variations in light intensity, amount of water vapour, changes in temperature, and other stimuli.

Recognition of the enormous amount of water evaporated from the leaf surface, water which must have been pumped out of the soil, led botanists to enquire as to the pathway of the transpiration current and the nature of the forces concerned in the ascent of the water to heights that could not be accounted for either by " root pressure " or the suctional effect of transpiration—the *vis a tergo* and the *vis a fronte* of the earlier physiologists; but this, perhaps one of the most hotly debated questions in the whole range of plant physiology, I must leave over to the next lecture.

LECTURE IX

THE ASCENT OF SAP

Ever since the time of Hales's classic experiment on "ringing," physiologists have been unanimous in believing that the path of ascent of sap in plants is by way of the xylem, but how the sap is raised to the top of a lofty tree is a problem on which the most divergent views have been expressed. Hales's experiment, as one of our most recent authorities on the subject has rightly pointed out, was not aimed at determining the *path of ascent*, but at showing that there did not exist in plants any regular circulation such as occurs in animals. Hales himself thought that ascent took place in the wood vessels in virtue of capillarity, while Christian Wolff believed that expansion of air co-operated in the upward movement. Both these agencies were, however, soon seen to be inadequate to explain the ascent beyond a certain height, nor did any greater success attend the comparison of the conditions existing in the wood with a Jamin's chain of water columns and air bubbles.

The first of the newer attempts to explain the movement on physical principles was that made by Sachs, in 1879, in a paper on "The Porosity of Wood," and, in 1882, in his *Lectures on the Physiology of Plants*. "The wood," he says, "owes its significance as the organ for conducting water to a series of most remarkable properties, which are found in no other natural body. It depends not upon a phenomenon of capillarity, but upon imbibition and swelling. The one point of special importance to be considered here is the facile mobility of the water thus held fast in the cell walls. The ascending current of water depends upon the motion of the relatively small number of water molecules which are contained between the micellae of the wood cell walls. This much is established, that this movement can only occur when the wood cell walls at the upper end of this system lose a portion of their water molecules. By this loss its state of saturation with water becomes disturbed, and the equilibrium altered; the parts of the wood cell walls which have become poorer in water will tend to restore the equilibrium by attracting water from the nearest wood cells, which, in their turn and for the same reason, take it up again from parts of the wood situated lower, until finally this movement, extending backwards, proceeds from the foliage of a land plant down through the stem into the young roots, which absorb the water out of the earth." You will see from this extract that Sachs attributes the whole ascent to the suctional effect of transpiration acting on molecules of water travelling in the walls of the xylem elements.

Apart from the fact that the anatomical features presented by the xylem vessels and fibres all tend to throw grave doubts on the probability of Sachs's imbibition theory, the obvious method of directly testing its truth was by occluding the lumina of the vessels and observing whether or not the leaves wilted, and it seems strange that Sachs did not apply that test himself. This was done by Elfving and Vesque in 1882 and 1883 respectively, and also by Kohl in 1885 and by Strasburger in 1891, using such occluding materials as cacao butter, melted paraffin, or gelatine. These experiments proved conclusively that Sachs's theory was incompatible with the facts, for the results showed that although some small quantity of water might ascend by the walls of the vessels, on the occlusion of the lumina, the amount so raised was quite negligible.

All the mechanical theories of ascent having broken down under the test of experiment, it was left for Godlewski, in 1884, to suggest an explanation based on the osmotic activities of living cells. "He assumed a periodic change in the permeability of the osmotic membranes of the parenchymatous cells contained within the wood in order to bring about a pumping action which would account for the raising of water in the tracheae of the stem. Thus, supposing a cell of a medullary ray in contact with eight tracheae, four on each side, to draw water into itself and to increase its turgor so that its protoplasmic membrane is considerably stretched, and assuming the osmotic pressure of the cell and the resistance to filtration of the membrane opposite to one trachea to be periodically and suddenly diminished owing to a chemical change, then it is evident that the contractility of the protoplasm will cause water to escape through the most permeable spot of the membrane, viz. into the trachea opposite to which filtration is most easy. Once in the trachea, Godlewski assumed it to move upwards until it was drawn into a medullary ray cell lying at a higher level in the stem. The reason given for the motion upwards in the trachea rather than downwards in obedience to the gravitational force, is because the air pressure in the tracheae above is less than in those at lower levels. . . . Godlewski claimed for his hypothesis that it explained the relation of the tracheae to the parenchymatous tissues, the radial position of the bordered pits, which facilitates a staircase motion of the water upwards in the stem, and the radial intercellular spaces along the medullary rays, which afford the aeration necessary for the respiratory liberation of energy in these cells" (Dixon).

In the same year Westermaier put forward the same hypothesis, though in a somewhat altered form, the parenchyma being regarded as the chief pathway of ascent and the vessels being rather water reservoirs than conduits.

I need not mention more than one critical experiment which aimed at disproving the vitalistic theory, viz. that of Strasburger, who showed that stems more than ten metres in height still continued to draw up water after all the living cells in it had been killed by exposure to a temperature of 90° C.

In 1894 an entirely new explanation of the puzzling phenomenon was put forward by Dixon and Joly, and in the following year by Askenasy. The theory is based on the cohesion of water molecules or the property possessed by water of resisting tensile stress, so that a column of liquid might be lifted bodily up a narrow tube; as Dixon puts it, " the water in the conducting tracts of high trees hangs there by virtue of its cohesion."

As I am not attempting to lecture to you on plant physiology but only on the historical growth of our knowledge of botanical problems, you will not expect me to recount the experimental evidence brought forward by Dixon and Joly in support of their hypothesis. That you must read for yourselves, and perhaps most conveniently in Dixon's monograph on the subject published in 1914. I will content myself with quoting the summary the author gives in his concluding chapter.

"The transpiration stream is raised by secretory actions taking place in the leaf cells, or by evaporation or capillarity (imbibition) at their surfaces drawing water from the tracheae. The state of saturation surrounding these cells determines which of these agencies is effective.

"The configuration, physical properties, and structure of the wood render the conducting tracts of plants highly inefficient if regarded as a system for conveying water urged upwards by pressure or drawn upwards in the substance of the woody walls. The distribution of living cells in these tracts is such that their actions cannot account for the rise of water observed, and there is no reason to believe that the elimination of these activities, if attended by no secondary changes in the conducting tracts or transpiring leaves, will arrest the transpiration stream.

"While thus structural and physiological evidence prevent us from accepting any of the previous physical or vital theories, the same configuration, physical properties, and structure of the wood compel us to admit that the water in the conducting tracts, when not acted upon by a *vis a tergo*, must pass into a state of tension. This state is necessitated by the physical properties of water when contained in a completely wetted, rigid, and permeable substance which is divided into compartments. Therefore when root pressure is not acting and when leaves of trees are transpiring, the cohesion of their sap explains fully the transmission of the tension downwards, and consequently explains the rise of sap.

"Resistance to a current of water moving through wood at the velocity of the transpiration stream is approximately equivalent to a head of water equal in length to the wood traversed. Hence the tension applied to the upper end of the water columns, which will be able to raise the transpiration stream in a tree, must equal the pressure produced by a head of water twice the height of the tree. In a tree 100 m. high, therefore, a tension of 20 atm. must be produced.

"The cohesion of sap amounting, as it does, to at least 200 atm. is in no way taxed by this tension.

"The transpiring cells of the mesophyll normally remain turgid during transpiration; accordingly we would expect, if our line of reasoning is correct, that in high trees the osmotic pressure keeping them distended must correspond in magnitude to the tension necessary to raise the sap.

"This surmise has been confirmed by determinations of the osmotic pressures of the saps of various leaves. These pressures have always been found adequate to resist the transpiration tension; but in many cases other factors enter in, and the pressures developed are much in excess of those demanded by transpiration.

"Finally, it has been shown that the stored energy set free by respiration in leaves is quite sufficient to do the work of secretion against the resistance of the transpiration stream; while, when the vapour pressure of water in the surrounding space is low, and when evaporation is doing the work of raising the sap, the expenditure of energy in this process will reduce the quantity of water evaporated only by an imperceptible amount."

The Absorption of Minerals

The absorption of water by the roots and its transport to the leaves naturally leads me to speak of the nature of the minerals dissolved in it and carried by it in its upward passage. There was no dearth of papers on the subject of plant ash and the functions of the ash constituents during the latter half of the nineteenth century, but it is astonishing how little was added to our knowledge in spite of all the labour expended.

Physiologists started from De Saussure's law, viz. that for every salt capable of absorption there was a certain concentration at which the root hair took in the solution as it was presented to it; if the concentration exceeded that optimum the plant absorbed less salt and more water. Subsequent investigators, however, soon discovered that the law was by no means of universal applicability. It was found that when a plant was cultivated in a medium composed of many salts in different proportions, it did not absorb them equally and indifferently, and that roots of different plants grown in the same soil took up the salts in different proportions. This led to the belief that plants had a power of selection, both qualitative and quantitative. But this view also was somewhat discounted by the discovery that plants were able to absorb not only salts of no service in metabolism but even some that were distinctly injurious.

Liebig had expressed the view that absorption of minerals was due to the solvent action of roots, a conception that seemed to be supported by Sachs's statement in 1865 that roots were able to corrode polished slabs of marble, a phenomenon which was attributed to the exosmosis of certain organic acids by the roots. By the end of the century this idea had been completely exploded by Czapek's work published in 1896.

In all the enquiries made during the later years of the century as to the functions of the different soil constituents two methods of investigation had been followed. One was that introduced by Salm-Horstmar in 1860, viz. to compound an artificial soil of insoluble constituents and then to add to it solutions of soluble salts known to be of service in plant nutrition; the other was that introduced by Sachs and Knop in 1860, and improved by Nobbe in 1862, known to you as the "water-culture" method. The results obtained seemed to show that, in addition to carbon, hydrogen, oxygen, and nitrogen, the metals potassium, calcium, magnesium, iron, manganese, and the non-metals phosphorus, sulphur, silicon, and, doubtfully, chlorine, were essential to all green plants. Salm-Horstmar affirmed that sodium was not an essential ash constituent. Salm's results were in the main confirmed by Birner and Lucanus in 1866, save that they claimed that sodium was essential while silicon and manganese were not.

I think it is scarcely worth while to recapitulate all the attempts that were made to attribute a special function to each element, for no sooner was a statement made by one investigator than it was contradicted by another. Take iron for instance. Gris, in 1843, showed that this metal was essential to the formation of chlorophyll. This led to the belief that iron was a constituent of the pigment, and some chemical physiologists, *e.g*

Hansen, claimed to have demonstrated its presence, while others, such as Gautier and Molisch, denied that it occurred in the analysis of chlorophyll, a view that is now accepted, although it is impossible to say in what way the iron is really essential. I might quote to you similar differences of opinion in relation to the functions of every one of the metals and non-metals I have mentioned. Perhaps the truth is that all the "elements of the ash play many parts, all co-ordinated by the living substance of the organism."

The Absorption of Nitrogen

You will doubtless remember that in 1861 Boussingault succeeded in demonstrating that the atmosphere was not the source of nitrogen to the plant, but that all of that element employed in metabolism was obtained from salts of nitric acid and ammonia in the soil. Experiments similar in their nature to those carried out by the French physiologist were conducted by Lawes and Gilbert in England. Lawes, in the early years of the century, had been investigating the values of certain manures at Rothamsted in Hertfordshire, and had acquired a considerable fortune from his patents for the production of superphosphates. In 1843 Lawes invited Gilbert, then a young man of twenty-six, who had studied chemistry at Giessen under Liebig, to join him, and so began a scientific partnership that lasted for fifty-eight years. More than one hundred monographs were published from the Rothamsted Institute between 1843 and 1901 in the joint names of the two investigators. The partnership was an ideal one, for Lawes was the practical man with a wide knowledge of the needs of agriculture, and Gilbert possessed the scientific brain, trained in a famous chemical laboratory, that could work out the best method of achieving the result aimed at.

Before describing to you the fundamental advances

made by Lawes and Gilbert it may be well that I should remind you of the views of Liebig on the relation of plants to the soil. Every crop, said Liebig, demanded and took from the soil certain minerals, and the essential mineral present in minimum quantity determined the success or otherwise of the crop in question. Consequently the purpose of manuring is to supply the deficiency in minerals, for both the carbon and the nitrogen could be obtained in abundance from the atmosphere. After a few experiments had been made, Gilbert dissented from this view, and found himself compelled to support Boussingault. An extensive monograph on the subject appeared in 1861 in the *Philosophical Transactions of the Royal Society* entitled, "The Sources of the Nitrogen of Vegetation; with special reference to the question whether Plants assimilate free or combined Nitrogen." In this paper Lawes and Gilbert state as the result of their experiments that "we have in no case found any evidence of an assimilation of free or uncombined nitrogen." This statement entirely supported Boussingault's conclusions, and so, as you will readily understand, it was—to use an Americanism—"up against" the followers of Liebig to prove their point; Lawes and Gilbert were not called upon to prove the negative, though they furnished abundant evidence in support of the correctness of their position.

In the same paper there is another almost side-note which opened up a new aspect of the whole problem of nitrogen assimilation. Let me give it you in full. "In our experiments with leguminous plants the growth was less satisfactory, and the range of conditions possibly favourable for the assimilation of free nitrogen was, therefore, more limited. But the results recorded for these plants, so far as they go, do not indicate any assimilation of free nitrogen. Since however in practice leguminous crops assimilate from some source so very much more nitrogen than graminaceous ones under ostensibly equal circumstances of supply of combined nitrogen, it is desirable that the evidence of further experiments with these plants under conditions of more healthy growth should be obtained."

In Valerius Cordus's *Historia Plantarum* this sentence occurs: "The root of the lupine is slender, woody, white and without useful properties, parted into a few slender fibres upon which there grow small tubercles." These words were written three centuries before the Rothamsted researches were published, and it is strange that no one during all these years made any effort to explain why leguminous plants only should possess these, seemingly, pathological swellings.

The next step in advance was taken by the French chemist Berthelot, who, in 1876, drew attention to the fact that, although plants did not absorb free nitrogen from the air directly, a constant fixation of atmospheric nitrogen was taking place in the soil and also in the air under the influence of silent electric discharge. In 1885 he associated the fixation of nitrogen in the soil with the presence of micro-organisms, and in the following year Hellriegel and Willfarth showed that the formation of nodules on the roots of Leguminosae was intimately connected with the absorption of nitrogen by such plants; so that at last, after three centuries of waiting, Cordus's discovery came into its own. These results Gilbert reluctantly accepted, even explaining his failure (in 1861) to demonstrate the fixation of nitrogen by Leguminosae by admitting that he had calcined the soil in which his experimental plants had been cultivated.

The root nodules were investigated in 1887 by Marshall Ward, who traced their origin to a microfungus which enters the root hair from the soil and causes hypertrophy of the cortical cells, and in the following year Beijerinck isolated the bacterium concerned and named it *Bacillus radicicola*. The whole life-history of the organism was finally worked out in 1899 by Miss Dawson, studying in Marshall Ward's laboratory.

Meanwhile several investigators had been following the line of research opened up by Berthelot on the fixation of free atmospheric nitrogen by bacteria in the soil, but I must leave you to study the details of these investigations in the textbooks, singling out perhaps one name, that of Winogradski, to whose labours we owe much of our knowledge of the remarkable chemical and biological activities of which the soil is the scene, all concerned in the presentation of combined nitrogen to the plant in such forms as may be most readily appropriated by it. Warington at Rothamsted also contributed substantially to the solution of this very difficult problem. One remarkable fact was made clear by his researches, viz. that the activity of the nitrifying organism is retarded by the presence of organic nutrients. Winogradski followed up this discovery by cultivating the bacteria in a sterilised solution of ammonium sulphate, potassium phosphate, and basic magnesium carbonate. In such a solution he found that, in a few days, all the ammonia had been replaced by nitrate and that a sediment was formed composed of bacteria in a zoogloea, which was capable of inducing rapid nitrification in any fluid containing ammonia. Winogradski also found that the nitrification proceeded in two stages, first, the formation of a nitrite and then of a nitrate. Further research showed that there were two organisms concerned, one transforming ammonia into nitrite and a second transforming nitrite into nitrate. Both of these obtained their carbon from the carbon dioxide of the air or from carbonates in the solution, but in neither case from organic compounds of carbon. Later on, in 1890, Winogradski showed that it was the carbon dioxide only that was the source of the carbon. This constructive metabolism was independent of light and, of course, of chlorophyll, and hence was termed chemosynthesis (in contrast to photosynthesis), and shown to be carried out by energy supplied by the oxidation of ammonia.

Winogradski also investigated the nutritive phenomena of so-called iron and sulphur bacteria. The latter were known to belong to the genus Beggiatoa, which were filamentous forms growing in water containing sulphuretted hydrogen. The energy in this case is obtained by oxidation of sulphur. Synthesis of organic material may thus take place without the presence of chlorophyll or light, and you will see how fundamentally important the study of these synthetic processes becomes in view of the speculations on the origin of life on the earth.

The Problem of Photosynthesis

During the past few years I have had occasion to collect all the references I could find connected with the subject of photosynthesis or carbon assimilation, from the earliest times to the present day. On looking through the volumes of notes I have accumulated I find I have the titles and abstracts of no less than 756 papers on the subject for the period 1860-1900. Quite half of these are probably of no importance, but I almost despair of doing anything like justice in the time at my disposal to the labour and energy represented by even a fraction of the remainder.

After 1859 the name of Sachs confronts us at every turn, and, whatever we may think of him as a historian, we must assign to him the place of honour as the first since De Saussure to revive the experimental side of plant physiology and submit the phenomena of living nature to critical laboratory investigation and analysis. His first contribution to the elucidation of the photosynthetic problem dealt with the influence of light on chlorophyll. He showed that the colourless plastid may form in darkness, but that light is essential to greening; that the light need only be feeble, and that intense light retards the development of the pigment. In 1860, Sachs, with the aid of an instrument he termed a diaphanoscope, estimated the degree of penetration of light through tissues of varied thickness, and showed that light did actually pass through structures that were opaque to the eye. He also determined which rays of the spectrum penetrated most deeply, and showed that the spectrum of the leaf was identical with that of an alcoholic solution of chlorophyll.

In the same year Frémy analysed the green pigment by shaking up an alcoholic solution with ether and hydrochloric acid and obtained a blue-green acid alcoholic solution and a yellow ether solution; the blue-green pigment he termed phyllocyanin and the yellow one phylloxanthin.

In 1862–63 Sachs published a series of important papers dealing with the intimate structure of the chloroplast and the origin of the starch grains in it, and showed that the appearance of the starch was associated with illumination; in his own words, "the starch in the chlorophyll body is a product of the living chlorophyll and it may be said in general that starch in the chlorophyll arises by the assimilatory capacity of the latter." From the leaves the starch is transported to wherever it is required, and travels in the form of glucose, dextrin, cane-sugar, etc., and may be redeposited as starch in other organs. He also discussed the chemical and physical properties of the pigment and confirmed the earlier observation that iron was essential to its formation.

In 1864 Stokes published a paper on the constitution of chlorophyll, which, in a very remarkable way, anticipated the work of the most recent chemical investigators of plant pigments, viz. Willstätter and his pupils. Stokes affirmed that chlorophyll was a mixture of two green and two yellow pigments, the former being fluorescent but the latter not. This was not only true of leaf green taken from land plants but also of the pigments extractable from Phaeophyceae, save that in these plants one of the green constituents is replaced by a body he called chlorofucin and one of the yellow by fucoxanthin. "The four substances," he adds, "are soluble in the same solvents and three of them are extremely easily decomposed by acids and even acid salts, such as bisoxalate of potash, but by proper treatment each may be obtained in a state of very approximate isolation so far at least as coloured substances are concerned." In a further paper published in the same year he emphasises the value of carbon disulphide in conjunction with alcohol, which enabled him "to disentangle the coloured substances which are mixed together in the green colouring matter of leaves."

Sachs advanced our knowledge of the subject still further, in 1864, by proving conclusively that starch disappeared from the leaf at night and reappeared by day, and disappeared even by day when the leaf was shaded. He investigated the effect of changes in temperature on greening, and determined the optimal temperature for the process in different plant types. He also asserted that chemical changes took place most vigorously when leaves were exposed to yellow-red rays; employing potassium bichromate and ammoniacal copper oxide as screens, he cultivated etiolated seedlings under the rays transmitted through these solutions, and compared the results with those obtained by the action of the same rays on photographic paper.

During the succeeding four years nothing very material was added to our knowledge of chlorophyll and its functions, but in 1868 two highly important papers appeared by Becquerel and by Boussingault. Becquerel made the first attempt to determine the energy relations of the leaf, and estimated that about $\frac{1}{250}$ part of the available solar energy was stored in the leaf in the potential form, an estimate that suffered considerable revision in later years at the hands of such investigators as Timiriazeff, N. J. C. Müller, Detlefsen, and Brown and Escombe.

Boussingault, also in 1868, tried the effect of growing plants in an atmosphere of pure carbon dioxide, and in mixtures of that gas with neutral gases like nitrogen and hydrogen. He found that oxygen exhalation ceased when carbon dioxide alone was present, and attributed this result, not to the absence of oxygen, as De Saussure held, but to the density of carbon dioxide, seeing that assimilation recommenced as soon as the carbon dioxide was diluted by one of the neutral gases above mentioned. He also controverted De Saussure's statement that plants give off nitrogen during photosynthesis, and established the constancy of the volumes of carbon dioxide absorbed and oxygen exhaled. He further expressed the view that the pathway of gaseous exchange was not by the stomata but by the cuticle, a view held for many years, and one in which he was supported by Barthélemy, who held that stomata were valves for the regulation of the exit of gases and not for their entry.

In 1870 Baeyer published a paper which had a profound influence on the trend of research during the years to follow, right up indeed to the present day. His theory was based on Butlerow's discovery in 1861, that a substance possessing some of the properties of a sugar could be obtained by heating a condensation product of formaldehyde in an alkaline medium. Baeyer's hypothesis has been so often referred to in the text-books that I think it advisable that you should have it presented to you in his own words, as translated by Jörgensen and Stiles. "The general assumption in regard to the formation in the plant of sugars and related bodies is that in the green parts carbon dioxide under the action of light is reduced and by subsequent synthesis transformed into sugar. Intermediate steps have been sought for in organic acids—formic, oxalic, or tartaric—which may be regarded as reduction products of carbon dioxide. According to this view, when the green parts of plants are most strongly subjected to the action of the sun's rays, an accumulation of acids should take place and gradually be replaced by sugar. So far as I know this has never been observed, and when it is remembered that, in the plant, sugars and their anhydrides are formed under all circumstances, whereas the presence of acids varies according to the kind of plant, the particular part of it, and its age, then the opinion already often put forward that the sugar is formed directly from the carbon dioxide increases in probability. Butlerow's discovery provides the key and one may indeed wonder why it is that, so far, it has been so little utilised by plant physiologists. The similarity which exists between the blood pigment and plant chlorophyll has often been referred to and it is probable that chlorophyll as well as haemoglobin binds carbon dioxide. Now when sunlight strikes chlorophyll which is surrounded by carbon dioxide, the carbon dioxide appears to undergo the same dissociation, oxygen escapes and carbon monoxide remains bound to the chlorophyll. The simplest reduction of carbon monoxide is to the aldehyde of formic acid; it only requires to take up hydrogen, *i.e.* $CO + H_2 = HCOH$. This aldehyde is then transformed under the influence of the cell contents as well as by alkalies into sugar. As a matter of fact it would be difficult, according to the other view, by successive syntheses to reach the goal so easily. Glycerine could be formed by the condensation of three molecules and the subsequent reduction of the glycerine aldehyde so produced. The formation of sugar by a more complex method is of course not excluded, and it might be quite possible that plant acids under certain conditions are transformed into sugar which in a thousand different forms helps to build up the body of the plant. How the cell contents act in order to effect the condensation of formaldehyde cannot be settled *a priori*, but one may assume that the sugar so formed remains bound with the cell contents and later, according to circumstances, splits off as a carbohydrate or glucoside."

Although this hypothesis played so important a part in the subsequent interpretation of the scheme of plant nutrition, it is remarkable that Sachs makes no mention of it either in his textbooks or in his periodical publications. The hypothesis has undergone several modifications at the hands of later investigators, notably Reinke, Erlenmeyer, Crato, and Bach, and has been adversely criticised by many authors, but no more plausible explanation of the phenomena was forthcoming up to the end of the century.

The years 1869-71 heralded the entry of two new names into the ever-increasing army of workers on the photosynthetic problem, viz. Pfeffer and Timiriazeff. Pfeffer dealt more especially with the effect of individual rays of the spectrum in carbon assimilation. When Draper examined this aspect of the problem he was faced with two difficulties, viz. to obtain light of sufficient intensity and at the same time to maintain purity of colour. Sachs's double bell jars permitted the use of coloured solutions as light screens, but the light transmitted was not monochromatic. Pfeffer adopted the plan of using fluids which absorbed only one type of ray, and contrasted the results with those given by light that had passed through pure water. He concluded that each ray of the spectrum, whether combined or otherwise, had a specific power of decomposing carbon dioxide, and that the most effective rays were those in the yellow region of the spectrum, the maximum of carbon assimilation coinciding with the maximum of illumination, while the more refrangible rays in the blue violet region were comparatively unimportant and the dark heat rays were quite useless.

In the same year (1871) Lommel claimed that the greatest decomposition of carbon dioxide took place between Fraunhofer's lines B and C, and he was supported in this view by Müller. Timiriazeff's first contribution also dealt with the unreliability of Draper's results, for he could find no coincidence between the maximum of carbon dioxide decomposition and that of luminosity.

In 1872 Kraus published his well-known method of separating crude leaf green into cyanophyll and xanthophyll by shaking up an alcoholic solution of leaf green with an equal volume of benzin, but many of his critics insisted that chlorophyll was a single pigment, and that cyanophyll and xanthophyll were decomposition products.

Quite a number of papers appeared in 1873 and 1874 which aimed at confirming the fundamental thesis laid down by Sachs and at filling up the gaps in the story, while several further attempts were made to determine the exact composition of chlorophyll and to isolate it in a pure condition. In 1875 Timiriazeff attacked the views of Draper and Pfeffer more especially, and advanced new evidence in support of his previous conclusion that the really effective rays in photosynthesis were those between the lines B and C of the spectrum. He employed the eudiometric method of measurement of oxygen exhalation, discarding Dutrochet's and Sachs's gas bubble method as unreliable.

Among the many publications of the next five years I need only mention one by the French chemist Gautier, in which he showed that chlorophyll contains no iron, and that chemically the green colouring matter was allied to bilirubin, one of the bile pigments. In 1875 Hoppe-Seyler attempted a new analysis and confirmed the absence of iron but demonstrated the presence of magnesium.

The year 1879 saw the publication of the first of a long series of monographs by Pringsheim, in which he tried to establish an entirely new theory as to the function of chlorophyll. Although Pringsheim's hypothesis has long since been discarded, the papers themselves and the criticisms to which they gave rise occupy so much of the literature of the period that, from the historical point of view, I may be justified in spending a few moments on this remarkable theory.

With the aid of a lens and a heliostat Pringsheim exposed green tissue to highly concentrated light, and found that after a few minutes the chlorophyll was completely destroyed, although the protoplasm at first suffered little alteration. He convinced himself that this destructive effect was not due to high temperature and was dependent on the presence of oxygen. On the strength of these observations he suggested that chlorophyll operated as a protective screen, moderating the injurious effect of light on the protoplasm and, by its absorption of the so-called chemical rays, acting as a regulator of respiration. He also claimed to have discovered a new substance of a hydrocarbon nature which he called hypochlorin, which he thought was the first product of assimilation, and from which starch and other compounds were derived. This hypochlorin was obtained by heating green cells with dilute hydrochloric acid or treating them with steam; it appeared on the margins of the chloroplasts as small glutinous red-brown droplets or needles. It formed the oily basis of the plastid, and was soluble in alcohol, ether, and benzin, but insoluble in water. After separation it solidified into an obscurely crystalline resinous or waxy substance. Increase in starch, Pringsheim claimed, was accompanied by decrease in hypochlorin. Yellow seedlings contained no hypochlorin, but it at once made its appearance in light and as soon as the chlorophyll became evident. It did not, however, occur in the dark in the case of Gymnosperm seedlings, which develop chlorophyll in darkness. When hypochlorin was removed from the chloroplast there was left a spongy framework, the plastid proper. Thus the green pigment in absorbing light protected the hypochlorin from oxidation.

For five years or more paper after paper appeared on this subject, indeed much of the literature published in this branch of physiology between 1880 and 1885 was devoted either to expositions of Pringsheim's theory or to criticisms of his methods and refutations of his results, but it would serve no useful purpose to follow the matter further; we may leave Pringsheim's theory and his hypothetical first product of assimilation as, in Hansen's words, an "unsubstantial dream."

In 1880 A. F. W. Schimper produced an elaborate monograph on the structure and origin of starch grains from chloroplasts, in which he showed that they might also arise from what he termed "amyloplasts" in non-green organs, and these, he affirmed, became chloroplasts when exposed to light. In the following year Engelmann drew attention to the behaviour of the motile form of *Bacterium termo* when placed in the vicinity of an illuminated green cell. The movements of these organisms were particularly active under red and orange rays, very feeble under green, more vigorous under blue, but ceased under ultra-red rays. The movements were attributed to the extreme sensitiveness of the bacterium to the presence of the most minute traces of oxygen, and Engelmann suggested the employment of such motile bacteria as tests for the evolution of oxygen, and consequently of the intensity of photosynthesis in green cells placed in different regions of the spectrum. This method he elaborated in subsequent papers, and it has been extensively used by many investigators since his day.

Difficulties had been raised by several critics of Baeyer's theory, on the grounds, well justified as it would appear, that not only were carbon dioxide and water very intractable substances and difficult to decompose, but that carbon monoxide and free hydrogen could not be detected, even in the faintest traces, in the fresh leaf. Reinke met these objections, in 1881, by suggesting that it was not carbon dioxide and water that were decomposed but carbonic acid gas, which was a very unstable compound; by losing a molecule of oxygen it might become formaldehyde, afterwards converted by polymerisation into grape sugar.

Schmitz, in 1882, though he did not touch on the physiological problems concerned, gave a very full account of the structure and development of the chloroplasts, more especially of Algae. He expressed dissent from the view that the plastid formed a reticulum, and held that this was induced by the action of acids, and that the pigment was diffused through a homogeneous matrix. The chloroplasts, moreover, never arose *de novo* from the protoplasm, but always as a consequence of the division of pre-existing plastids. On this point he was supported by Schimper, who published a paper on the subject in the following year.

Another contribution to the subject came from the pen of Reinke in 1884, in which he discussed the much debated question as to the relative efficiency of the different rays of the spectrum in the exhalation of oxygen from green cells. In this paper he described an apparatus which he termed a spectrophore and which he employed in his researches. A pencil of light from a heliostat is projected through an objective on to a large prism, by which the spectrum is thrown on a screen consisting of two movable boards, arranged so as to leave a slit between them. By manipulating the boards the slit may be opened or reduced to any extent and any portion of the spectrum allowed to pass through. The rays selected fall on a biconvex lens behind the screen and are focussed on the object to be examined. Thus light of any colour or any combination of colours up to pure white may be obtained by varying the aperture of the slit. Reinke estimated the rate of exhalation of oxygen by the old gas-bubble method, and from a large number of experiments he concluded that the absolute maximum of evolution of oxygen is effected by rays of wave-length 690-680, *i.e.* those between Fraunhofer's lines B and C.

He found the curve of intensity of exhalation fell rapidly towards line A and also, but less steeply, towards E, and then slowly to H. He thus denied that there was a second maximum in the more refrangible part of the spectrum, the existence of which had been asserted by Engelmann.

Baeyer's hypothesis of 1870 led in the succeeding years to various experiments in feeding green plants on formaldehyde, the most successful of which were those of Bokorny, who claimed to have obtained a gain in weight in the green Alga, Spirogyra, after supplying it with various substances which yielded formaldehyde on decomposition. The chief argument used by Bokorny and others was that formaldehyde was not poisonous if supplied in minute quantities and under conditions when it could be at once polymerised and so rendered innocuous.

In 1893 a paper by Brown and Morris appeared which created quite a sensation in laboratories of plant physiology, and to which I must make a brief reference. The authors, while engaged on the determination of diastatic enzymes in leaves, came to the conclusion (as did also Gautier at an earlier date) that cane-sugar was the first carbohydrate formed in photosynthesis, and, further, that the larger part of the products formed in the synthetic process never assumed the form of starch.

You will remember that Boussingault, in the year 1868, concluded that the pathway of gaseous exchange was not by the stomata but by the cuticle of the leaf, and that Barthélemy supported him in this view. In 1894 F. F. Blackman, by aid of an ingenious piece of apparatus, re-investigated the whole question. His experiments led him to the conclusion that, under normal conditions, practically all the carbon dioxide passed through the stomata, both inwards and outwards, although when the stomata were blocked and the tension of the gas was sufficiently high, which it rarely was in nature, some might pass through the cuticle. He pointed out that the error into which Boussingault had fallen was due to his having used far too high percentages of carbon dioxide. Blackman's results thus confirmed and greatly extended the conclusions of Mangin, announced some ten years previously. Blackman also showed that the distribution of stomata on leaves of different plants corresponded practically exactly with the volumes of gas passing into and out of the leaf, and thus completely set at rest the controversy that had been in progress since Boussingault's day.

Another very important piece of work was carried out in 1900 on the same subject by Brown and Escombe, in which the authors showed that stomata, as a general rule, were placed at intervals apart which were about eight to ten times the average diameter of the stomatal opening, and that this distribution coincided with the conditions which must be fulfilled under the law of diffusion of gases through perforated membranes. Wiesner, in 1879, had made some observations of the same nature but without the finished detail that is so marked a feature of Brown and Escombe's work. I must leave you to read for yourselves this highly important and convincing research; to attempt to condense it would destroy the beautifully logical mode of presentation in which the authors' conclusions are worked out.

Those of you who are chemists, and are familiar with Roscoe and Schorlemmer's great *Treatise on Chemistry*, will notice that I have not referred to the many papers that appeared during these forty years on the chemistry of chlorophyll, and that I have omitted all mention of such well-known names as those of Marchlewski, Schunck, Tswett, Tschirch, and others who investigated the chemical composition of the green pigment by all conceivable methods, and with results of the most varied and often contradictory nature. This omission is intentional, for if I attempted to give you some idea of the labours of one investigator I should have to deal with the work of those who followed him and who either controverted or modified his conclusions, and that would entail an expenditure of far more time than I can spare in a summary as brief as the present. Besides, I shall have to sketch out the modern position of the subject later on in relation to the work of Willstätter and his pupils, and it would therefore seem unnecessary to follow in detail through the maze of controversial literature to which I have referred.

LECTURE X

METABOLISM AND GROWTH

IN past lectures I have tried to sketch out for you the development of our knowledge of what may be termed the preliminary stages in plant nutrition up to about the years 1890-1900, viz. the absorption and transport of water and minerals, the acquisition of nitrogen and carbon, and the construction of the first products of photosynthetic activity. I must now add a word or two on the attempts that were made previous to the dawn of the twentieth century on the general metabolic processes taking place in the plant.

Towards the end of the nineteenth century it had become accepted as a biological axiom that a plant was a living organism in which a most complex series of chemical changes were constantly taking place, described collectively as "metabolic," and that these changes might be arranged in two categories—constructive, or anabolic, viz. those connected with the upbuilding of organic matter and the storage of energy, and destructive, or katabolic, those concerned with the disintegration of organic compounds, the release of energy and the excretion of waste. The goal of anabolism was obviously the formation of protoplasm, and the outcome and accompaniment of katabolism was the manifestation of what we call "life." It was also recognised that the anabolic phenomena might be divided into two phases, the first concerned with the acquisition of the raw materials and their construction into those higher compounds to which the term "food-stuffs" might be legitimately applied; the second, the appropriation of these food-stuffs by the protoplasm, with the accompanying phenomena known to the animal physiologist as digestion and assimilation.

There has been much controversy over the terminology of these anabolic phases, but I need not trouble you with the arguments for and against the use of such words as "photosyntax," "photosynthesis," "carbon assimilation," and so on; the important thing is to recognise that these phases in plant nutrition exist, and, as I have already frequently pointed out to you, that the animal differs from the plant in starting its nutritive processes, so to speak, at a higher level than does the plant.

As to the details of the metabolic changes, between the primary acquisition of the raw materials and the development of the finished product, little was known up to the middle of the nineteenth century. Indeed as late as 1860 the general account given in the textbooks was that "crude sap," as it was called, ascended by way of the xylem of the root and stem, became "elaborated" in some mysterious way in the leaf when that organ was exposed to light and air, and that the "elaborated sap" then descended in the form of a kind of mucilage or gum, from which reserve materials were condensed and deposited in various situations.

The first step in the enquiry into metabolism in general was taken by Hanstein, who, in 1860, showed that the removal of all tissues external to the cambium was followed by the development of adventitious roots above the wound. The obvious deduction was that the descending "elaborated sap" provided materials required for the growth of these roots, *i.e.* manufactured plasta. Sachs in 1864, and Godlewski and Pfeffer in 1873, showed quite conclusively that the starch that accumulated in the leaf as a result of photosynthetic activity in light was removed in whole or in part not only during the night but also by day in the form of a sugar. The method of transformation of the starch into sugar was at once made the subject of investigation, with the result that the enzyme "diastase," identified in 1833 by Payen and Persoz in germinating barley, was discovered in the leaf, and further that the sugar formed by its action was maltose. Brown and Morris, in 1893, went further, and claimed that glucose and fructose were also produced, and they thought that these two sugars arose from cane-sugar as the result of the action of another enzyme, "invertase," cane-sugar being, according to the authors, the first sugar to be formed in the leaf.

The anabolism and transport of proteins were also studied by many workers during the period 1860–80. Chief among these was Pfeffer, who detected amino-compounds in the descending sap, and Vines, who determined the presence of proteolytic enzymes in the leaf. Another problem that exercised the minds of physiologists in the same years was the exact path by which these soluble carbohydrates and proteins travelled. Sachs at first thought that the cortex was responsible for the transport; Haberlandt suggested the laticiferous tubes; Schimper claimed that the sheath of the vascular bundles was the medium, and Strasburger, Kraus, and others attributed the duty of translocation to the sieve-tubes of the phloem. How translocation was actually brought about, and what were the effective forces concerned, remained quite in doubt up to the end of the century.

Meanwhile much work had been accomplished on the products themselves. To start with, Schimper in 1880, Meyer about 1885, and, eight years later, Brown and Morris worked out the whole history of starch as a reserve substance, its origin from chloroplasts and amyloplasts, its dissolution by enzymes, and its transportation and re-deposition. In 1895 Meyer summarised all that was known on the subject, and cleared up some controversial points left unsettled by his predecessors. Although in the great majority of cases the carbohydrate reserve was clearly seen to be in the form of starch grains, other carbohydrates were recorded as replacing starch, *e.g.* glycogen, inulin, and even cane-sugar.

Our knowledge was by no means so extensive as to the nitrogenous reserves, although one of these, viz. aleurone, had been the subject of enquiry by Hartig as far back as 1855. Such work as was accomplished was carried out chiefly on seeds, first by Hoppe-Seyler, and later by Vines and two American investigators, Chittenden and Osborne.

Sachs, in his early researches on the carbohydrates of the leaf, had noticed the occurrence of oil as a reserve, and studied its origin and fate. In spite of the chemical difficulties involved, he held that fats might arise from carbohydrates, and carbohydrates from fats. Naegeli and others, some years later, regarded the fats as direct products of the protoplasm, while Wakker and Zimmermann, towards the end of the century, determined the existence of bodies to which the name of elaioplasts was given, whose duty it was to secrete oil drops.

Other organic substances, such as glucosides, were also identified in plants towards the end of the century, and believed by many authorities to be of the nature of reserves.

Enzymes

As we have already seen, all reserve bodies had to be transformed from the resting, usually insoluble, condition into a soluble state before they could be transported from the seats of storage to the regions where they were required as constructive plasta, and to effect this transformation certain agents were necessary; these agents, as you know, are enzymes.

The first enzyme to be identified was diastase, which, as I have already stated, was obtained in 1833, by Payen and Persoz, from germinating barley. After the middle of the century, investigator after investigator tackled the problem of the mode of action of these mysterious bodies that appeared to be capable of inducing profound changes in other substances without themselves undergoing any alteration in the process. Although most work was carried out on the first enzyme to be recognised—diastase—it was not long before other similar bodies were discovered and investigated. Thus in 1887 Le Clerc du Sablon extracted from the artichoke an enzyme which dissolved inulin and which he termed "inulase"; in 1888 Marshall Ward isolated from a lily-disease fungus a substance able to dissolve cell walls, to which the name "cytase" was given, and its occurrence in seeds with sclerotic endosperm, such as is found in many Palms, was proved ten years later by Brown and Morris, Gardiner, Bourquelot and Hérissey, and others. Similarly the transformation of cane-sugar was found to be due to an enzyme called "invertase," and several other sugar-altering ferments were rapidly added to the list, such as maltase, glucase, lactase, etc. The glucosides also appeared to require the services of such enzymes as emulsin, myrosin, etc., while the proteins had their specific enzymes, such as trypsin and erepsin, and fats were acted on by lipases of various types. In short, it soon became apparent that enzymes played a most important part in all metabolic changes, and that for each reserve the protoplasm was able to call into existence an appropriate enzyme, when it became necessary to transform the reserve from a resting condition into a body capable of passing through cell walls.

Assimilation of Plasta

If enzyme action be a subject of investigation bristling with difficulties, assimilation, or the actual appropriation of plasta by the protoplasm, is a problem that, as yet, we have been quite unable to unravel. The subject is, all the same, a fascinating one, for until the riddle is solved we must despair of acquiring any real conception of what protoplasm itself is, and any clear idea of what is meant by the term " life."

Not a few biologists have indulged in speculations on the subject, such as the human physiologist Pflüger in 1875, Loew and Bokorny in 1892, and Verworn two years later, but as these speculations would be meaningless to you at the present stage in your studies, I do not think there is anything to be gained by recounting them. Emil Fischer's work on the constitution of proteins belongs to a somewhat later period, but perhaps I may make a very brief reference to it here. Fischer showed that proteins, animal as well as vegetable, may be split into amino-acids, after the elimination of water. An amino-acid in turn is composed of a grouping of nitrogen and hydrogen atoms—NH_2—with a grouping of carbon, oxygen, and hydrogen atoms—COOH—known as carboxyl. If water (H_2O) be eliminated there is left CO—NH, which chemists call a " peptide linkage." The various amino-acids are composed of one or more NH_2 groups united with one or more COOH groups, and Fischer was able, after elimination of the water molecules, to form artificially a large number of peptides containing two or more amino-acids, and the suggestion is that these bodies are stages in the construction of the complex we call protoplasm.

Katabolism

In the animal world the intaking of oxygen and the exhalation of carbon dioxide, commonly known as " respiration," had, from a very early date, been recognised as associated with the exhibition of vital phenomena and the expenditure of energy, but it was not until the earlier years of the nineteenth century that physiologists distinguished between the obvious external gaseous interchange and the actual chemical processes taking place in the tissues. The zoologist had, apparently, a quite straightforward problem to solve, for he had to consider only one gas entering the organism and one gas escaping from it. The problem presented to the botanist was a much more complex one. If you look back to Ingen-Housz's account of gaseous interchange between the air and the plant you will, I think, realise that he was far from having a clear conception of the meaning of the various phenomena he observed taking place. Even De Saussure uses the terms "inspiration" and "expiration" in relation to the entry of carbon dioxide and the exit of oxygen in the photosynthetic process. Considering the condition of chemical knowledge in these early days, it was not to be wondered at that physiologists should be puzzled by observing in plants one kind of "respiration" taking place by day and another by night. It was Dutrochet, so far as I can see, who first (in 1837) asserted that respiration was fundamentally identical in both the plant and the animal world, and yet Liebig, four years after Dutrochet, flatly denied the occurrence of respiration in plants. Sachs, in his famous textbook published in 1868, cleared up the whole situation, and showed that respiration was, as Dutrochet had stated, essentially the same in both sets of organisms, but that in the green plant, when exposed to light, the respiratory phenomena were masked by the much more vigorous inhalation of carbon dioxide and exhalation of oxygen.

As soon as this fundamental thesis had been established, enquiry began to be made as to what exactly took place in the tissues—how was the actual oxidation effected? Working on the idea that respiration was essentially an oxidation of carbon—a kind of slow combustion in fact—investigators from about 1870 onwards poured out paper after paper, giving data as to the amounts of oxygen inhaled and carbon dioxide exhaled under different conditions of temperature, food-stuffs, and so on. The animal physiologists were also busy with the same problem about the same time, and their investigations showed that the exhalation of carbon dioxide must be distinguished from the inhalation of oxygen, since an animal such as a frog could live and give off carbon dioxide for some hours in the absence of oxygen gas. Conversely, De Saussure, in 1804, found that a cactus, though supplied with oxygen, gave off no carbon dioxide. De Saussure's experiments were reinvestigated by Dehérain in 1874, with the result that oxalic acid was discovered in large amounts in the tissues. It was obvious that, in this case at all events, respiration was not a simple combustion of carbon.

The trend of research, in short, went to show once more that the whole performance was intimately connected with the protoplasm and the instability of its constituents. In 1875 Pflüger had put forward a hypothesis that protoplasm had the power of auto-decomposition; that oxygen became incorporated in it intramolecularly and so increased its "lability" or instability, the final result being an "explosion." Detmer, in 1883, adopted this idea and suggested "that the primary source of the energy developed in living organisms consists in the splitting up of highly complex labile compounds, the decomposition products of which subsequently become oxidised, carbon dioxide and water being the ultimate products."

These ideas received considerable support from the researches of Pasteur on fermentation. As you already know, he had shown that many yeasts and other micro-organisms could exist in the absence of free oxygen, and indeed that several could not exist in its presence; these forms he described as "anaerobic." It was also shown by Hoppe-Seyler, in 1887, that the products due to the activities of putrefactive bacteria differed according as to whether the putrefactive process went on in the presence or absence of oxygen. In the former case oxygen, carbon dioxide, water, and ammonia were formed;

in the latter, hydrogen, marsh gas, leucin, and tyrosin. Higher plants also appear to have the power, in absence of free oxygen gas, of decomposing sugar, with the formation of alcohol and carbon dioxide, and these and similar phenomena led to the linking up of respiration with fermentation as was suggested by Pfeffer in 1885. This physiologist held that sugar was first of all decomposed into alcohol and carbon dioxide without the aid of oxygen, but that, when that gas was supplied, the alcohol was further broken up into carbon dioxide and water. To follow out the story I have thus so very briefly outlined would take me far beyond the limits I have prescribed for myself; besides, the subject becomes more and more of the nature of a problem in organic chemistry, and hence outside the boundaries of botany in the strict sense of the term.

Growth

I must now devote a sentence or two to the work that had been accomplished in a branch of physiological investigation which I have not as yet touched upon at all, viz. growth.

Our knowledge of the phenomena of growth really dates from the year 1873, when Sachs postulated that every cell, tissue, and organ passed through certain phases which he termed collectively the "grand period of growth." The first phase of growth was slow, but the rate gradually increased until the maximum activity was reached, when there followed a gradual decline to zero. Examples of this succession of phases will occur to you in abundance from your knowledge of the development of the root and stem from their apical meristems backward to the zones where extension ceases and the tissues assume their permanent size and shape. Sachs not only illustrated the "grand period" in great detail, but studied the variations that exhibited themselves in the regularity of the curve, as a consequence of alterations in the conditions of the environment, and associated growth with the state of turgor in the growing cells.

The investigation of growth phenomena was greatly facilitated by the use of the growth-lever, first employed by Baranetski in 1879, and afterwards developed into the self-recording auxanometer which you have so frequently seen in operation in laboratory experiments. To Sachs's papers, as also to those of Kraus, published between 1865 and 1870, we owe as well the foundations of our knowledge of the distribution of tensions in different growing tissues, and the consequent results in the maintenance of rigidity in succulent organs.

Some Advances in Botanical Knowledge since 1900

I propose now to deal with a few of the many special problems that have been the subject of botanical research since the beginning of the twentieth century. Some of these in their earlier phases I have already in some measure discussed, such as the evolutión of the conception of a phylogenetic classification of plants or the photosynthetic problem, but there are many others that I have scarcely even mentioned. In summarising some of these newer aspects of the science I must bear in mind, however, the warning uttered by Sir Walter Raleigh, who, in his *History of the World*, apologises for not writing the history of his own time—" which would," he says, " have been more pleasing to the reader "—on the ground that " whosoever in writing a modern history shall follow truth too near the heels, it may haply strike out his teeth." Having no desire to suffer this facial injury I do not propose to offer you a complete statement on any one of these subjects ; all I can do is to select some of what appear to me to be the most important lines of development of the science, leaving the lesser bypaths for your own exploration.

Mendelism and Evolution

You may remember one curious misconception the early botanists had in connection with the phenomena of reproduction. Even Schleiden thought that the plant embryo was carried by the pollen-tube to the ovule, whose chief duty it was to form a nidus in which the embryo might be properly fed and protected. The male was the source of the "seed," the female was the soil in which it was sown. This idea accounts for the use of the word "seed" in two senses—the ancient one, as signifying the germ of a new generation that always came from the male (*conf.* "The seed of David" and the Bible *passim*), and the modern one, indicating an embryo plant with its protective coverings. It was only after the invention of the microscope and the discovery of egg cells and fertilising cells that it was appreciated that the embryo was the product of fusion of constituent units from two individuals, or from two different parts of the same individual. Both of these units were of microscopic dimensions, although the female ovum was, as a rule, many times the size of the male sperm. Minute as these units were, the marvellous fact remained that, in some way or another, the characteristics of the parents were reproduced in the offspring. That "like tends to beget like" was an ancient truism, and it signified that the multifarious characters of the parents, both material and immaterial, were transmissible to their offspring, *i.e.* were inherited. This inheritance of parental characters did not exclude the appearance of individual variations in the offspring, nor indeed the resuscitation of features of some far-back ancestor, a phenomenon known as "atavism."

Of course the only possible agents by which the parental or ancestral characters could be transmitted were the germ cells, and when the microscope had become sufficiently perfected to reveal the presence of chromosomes in the nucleus and to enable us to follow the remarkable changes they underwent in cell division, it was but natural that the chromosomes should be regarded as in some way connected with the transmission. Each germ cell also was seen to contain the same number of chromosomes—one-half the number present in the nuclei of the adult—so that each parent contributed the same number of "bearers of parental characters" to the nucleus of the zygote.

I need not go into the theories advanced by earlier writers on this interesting subject; all of them were essentially based on the idea that the various organs of the body contributed certain units which were accumulated in the germ cells, ready to reproduce in the embryo all the structures of the parents from which they were derived. This conception, under the name of "pangenesis," was elaborated most fully by Darwin in 1868, who spoke of these hypothetical units as "gemmules." Darwin's theories were sharply criticised in the succeeding years by Galton, Brooks, De Vries, and others, but we need not spend time over these suggested modifications of pangenesis in view of the hypothesis put forward by Weismann in his work on *The Germ Plasm* in 1885. One conclusion in Galton's work should, however, be referred to, viz. the so-called "*law of ancestral heredity.*" According to this law, if the total inheritance transmitted to an individual be regarded as unity, then one-half is derived from the two immediate parents, one-quarter from the four grand-parents, one-eighth from the eight great-grand-parents, one-sixteenth from the sixteen great-great-grand-parents, and so on. This statistical result was obtained by Galton from a study of the transmission of colour in the fur of a breed of hounds.

The hypothesis usually known as "Weismann's theory of heredity" was really enunciated ten years previously by Jaeger, who wrote, "through a great series of generations the germinal protoplasm retains its specific properties, dividing in development into a portion out of which the individual is built up, and a portion which is reserved to form the reproductive material of the mature offspring." Unfortunately for the theory as put forward by Jaeger, embryological investigations failed to show that this supposed continuity of the germinal cells actually existed save in a very few instances. Weismann's adaptation of Jaeger's theory aimed at getting over this crucial difficulty in the following way. He held that a certain part of the fertilised ovum is, so to speak, put on one side from the very commencement of the developmental process to serve as a starting-point for the germ cells of the new organism, and to this hypothetical substance he gave the name of "germplasm." The germplasm is a part of the nucleus of extremely complex structure which has the power of growing and yet of retaining its primitive characters unaltered. Part of it, however, is changed into the nuclei of the ordinary somatic cells while retaining enough of its original efficiency to start a new organism asexually if need be. As Galton puts it, according to the Weismannian hypothesis, "the main line may be rudely likened to the chain of a necklace and the personalities to pendants attached to the links."

One important consequence of Weismann's work was to call in question the validity of the view that any modification of the individual induced by external conditions during its lifetime, in other words, an "acquired character," would be transmitted to the offspring. If, as Weismann thought, germplasm and somatoplasm were initially distinct, and if the germplasm was alone concerned in the transmission, it was obvious that modifications of the somatoplasm could not be transmitted, that any structural change in a part of the body induced by use or disuse, or by environmental or nutritive influences generally, never affects the germplasm in such a way as to cause the offspring to exhibit the modification that the parent had acquired, or even to show a tendency in that direction.

Ten years after the publication of Weismann's disturbing theories one of our chief English authorities on such problems, Bateson (from whose writings I am freely quoting) drew attention to the fact that species do not merge gradually into each other, but that there is always more or less marked discontinuity. How can this discontinuity be accounted for if the process by which species have arisen is one of accumulation of minute and almost imperceptible differences? Why are not intermediates of all sorts more abundantly produced in nature than is actually known to be the case? These criticisms seemed to be met to some extent by De Vries, who, in 1901, published an important work on what he termed *Mutations*. From a long study of Oenothera, the "evening primrose," he came to the conclusion that new varieties might spring into being at a bound, so to speak, and his work raised the suspicion that variation might be discontinuous and not due to the slow accumulation of infinitesimal additions. Were such a view accepted, it followed that it would be necessary to alter entirely our views on the doctrine of evolution as propounded by Darwin in 1859.

It was at this time, when our notions on heredity and variation had been so rudely shaken and thrown once more into the melting-pot, when, as Bateson says, "in the study of evolution, progress had well-nigh stopped, when the more vigorous, perhaps also the more prudent, had left this field of science to labour in others where the harvest is less precarious or the yield more immediate, when of those who remained some still struggled to push towards truth through the jungle of phenomena, whilst most were content supinely to rest in the great clearing Darwin made long since"—it was at this moment that a name burst upon the biological world that has now become familiar to every student of the science—the name of Mendel.

Gregor Johann Mendel was born in 1822 in Austrian

Silesia and, in 1843, entered the Augustine seminary of Altbrünn, subsequently proceeding to the University of Vienna, where he studied science. In 1853 he returned to Brünn as a teacher and finally as Abbot of the Monastery, where he remained till his death in 1884.

During the long quiet years of his residence at Brünn he seems to have taken a keen interest in the local scientific society and published several papers in its Journal, and also carried out a lengthy series of experiments in the Monastery garden on the hybridisation of peas and hawkweeds, an account of which he published in 1865 in the Journal of the Brünn Society. Whether it was because naturalists did not attach any particular importance to the Abbot's work, or because the paper itself was hidden in an out-of-the-way publication, the fact remains that these, now classic, experiments and the epoch-making conclusions drawn from them lay unknown for thirty-five years until, in 1900, the paper was discovered by De Vries, Correns, and Tschermak, and, as expounded by these well-known authorities, immediately attracted the attention of the whole biological world. Let us see what Mendel actually found and what conclusions he drew from his observations.

For the purposes of his experiments he selected the common garden pea which, in cultivation, exhibits quite a number of distinct breeds or strains. Some are tall, some are dwarf, some red flowered, some white flowered, some have smooth testas and some have wrinkled testas, and so on, and each of these strains, if self-fertilised, "breed true." Mendel then selected two strains that showed a marked contrast in any one of these characters, such as colour, height, surface of testa, etc., and watched the effect of cross-fertilisation of contrasted strains and afterwards of self-fertilisation of the resulting hybrids. In one case he selected height, and artificially crossed a tall strain with a dwarf one, and obtained the unexpected result that all the progeny were tall, no dwarfs and no intermediates appearing. Obviously, tallness was a character that overpowered dwarfness, so to speak, so he termed the character tallness "dominant" and dwarfness "recessive." He then cultivated the seeds resulting from the self-fertilisation of all these tall hybrids, but, instead of obtaining a second generation of tall plants, he got a mixture of tall and dwarf, again without any intermediates. After a number of experiments he discovered a remarkable numerical relation between the tall and the dwarf plants, viz. the former were found to be on the average three times as numerous as the latter.

The next step was to self-fertilise all the offspring of the hybrids of the second generation and collect and grow their seed in turn. The result was again unexpected. The seeds of the dwarf plants developed into dwarf plants and continued to breed true dwarfs in subsequent generations. Further, one-third of the tall plants also continued to breed true. The seeds from the remainder of the tall plants, however, behaved exactly in the same way as the tall plants of the first generation when self-fertilised, *i.e.* three were tall to every one dwarf. Expressed numerically, in the second generation out of every hundred plants 75 were tall and 25 were dwarf, and out of the 75 tall forms 25 bred true, so that as a final result we have 25 tall breeding true, 25 dwarf breeding true, and 50 tall not all breeding true but prepared to breed in the same proportion as the original first generations of hybrids.

The results of Mendel's experiments have been summarised by Professor Punnett in the following words: "In every case where the inheritance of an alternative pair of characters was concerned the effect of the cross in successive generations was to produce three and only three different sorts of individuals, viz. dominants which bred true, dominants which gave both dominant and recessive offspring in the ratio of three to one, and recessives which always bred true. Having determined a general scheme of inheritance, which experiment showed to hold good for each of the seven pairs of alternative characters with which he worked, Mendel set himself to providing a theoretical interpretation of this scheme which, as he clearly realised, must be in terms of germ cells. He conceived of the gametes as bearers of something capable of giving rise to the characters of the plant, but he regarded any individual gamete as being able to carry one and one only of an alternative pair of characters. A given gamete could carry tallness or dwarfness, but not both. The two were mutually exclusive so far as the gamete was concerned. It must be pure for one or the other of such a pair, and this conception of the purity of the gametes is the most essential part of Mendel's theory."

Let me now attempt to show you how far Mendel's theoretical explanation was in harmony with his experimental results, and in doing so I shall follow, but at the same time condense, the admirable sketch of the whole subject by the author from whom I have just quoted.

Since they bred true, the original tall and dwarf parents gave off respectively gametes with tall and with dwarf characters. When these gametes fused the zygotes naturally carried both characters; but since tallness is dominant the new plants will all be tall, although half the gametes produced by any one of them will have the tall character and the other half the dwarf. Let us suppose that the plant has 100 ovules, 50 with the tall character and 50 with the dwarf. Similarly, if it has 100 pollen grains, 50 will have the tall and 50 the dwarf character. Each ovule with the tall character has an equal chance of being fertilised by a tall or a dwarf pollen grain, and each ovule with the dwarf character has similarly an equal chance of being fertilised by a dwarf or a tall pollen grain. Thus on an average 25 tall ovules will be fertilised by 25 tall pollen grains, and

25 dwarf ovules by 25 dwarf pollen grains, and in both cases the products of the resulting zygotes (homozygotes) will breed true, for in the first 25 zygotes there is no dwarfness, and in the second 25 there is no tallness. The remaining 50 zygotes, however, have tallness and dwarfness combined (heterozygotes), but, since tallness is the dominant character, all the plants produced will be tall. Thus on theoretical grounds we ought to get 25 true talls, 50 dominant talls, and 25 true dwarfs, which is exactly what we do get from experiment. Further, the 50 dominant talls when bred from must each give the same results as were given by the original pair, and so on for each successive generation.

When plants with two or more alternative characters were experimented with, Mendel found that, although each pair of characters followed the law I have just stated, "the inheritance of each pair was absolutely independent of the other."

How, you may next ask, will this affect our conception of variation and the doctrine of evolution as a whole? That species vary greatly is well known to you, and you have only to look at any English Flora to see recorded often a large number of varieties under the species headings, *e.g.* the willows, brambles, buttercups, and roses. But if the number of "factors" or "unit characters" is comparatively small, how is this great variation to be accounted for? Let us see. If there be only one factor there can be only two possible forms, but if there be ten, then there may be 2^{10}, or 1024 possible forms; where the heterozygous and homozygous forms differ in appearance there will be three possible forms for each factor, and if there be 20 such factors the possible number of different individuals will be 3^{20}, or 3,486,784,401, hence the range of individual variation may be enormous, and, provided that the constitution of the gametes is unchanged and the respective factors are clearly defined, each variation concerned will be transmitted in obedience to the Mendelian law. External conditions do affect individuals in a most marked manner, but there is no evidence that these changes in the environment produce any change in the gametes formed by such individuals. Since the publication of De Vries's *Mutations-Theorie* in 1901, it has been customary to describe variations due to specific factors as "mutations," while those which are due to environmental influences are termed "fluctuations"; the former are inherited, the latter not.

If you will now recall our discussion of the theory outlined by Darwin in the *Origin of Species*, you will see how these new discoveries have altered our whole outlook on the evolutionary hypothesis as set forth in that epoch-making work. It is worth remembering that Mendel's work was published only five years after the *Origin* left the printer's hands, and one is tempted to speculate what would have been the form and fate of that famous book had Darwin been acquainted with Mendel's *Experiments on Plant-Hybridisation*. Perhaps, as Bateson says, "the history of the evolutionary philosophy would have been very different from that which we have witnessed." Strangely enough, on the other hand, Mendel, in his paper published in 1865, makes no reference to Darwin, although he does speak of "the spirit of the Darwinian teaching," in his brief note on "Hieracium hybrids" published in 1870.

It will be useful to recapitulate at this point the "natural selection" theory, so that the alteration in point of view may be clear to you.

The offspring of an organism inherits the fundamental characteristics of its parents and yet no offspring resembles its parent in every particular; it occasionally shows features that recall characters of its grandparents or even of some farther back ancestor, and it also presents individual idiosyncrasies of its own. Some of these variations may be of such a kind as to lessen its chances of success in the struggle for existence before it; some, on the other hand, may be entirely in its favour. Those individuals possessing any variations giving them a superiority, however slight, over their fellows will, on the whole, be more likely to survive than those which have not the variation in question. They will in this way be "naturally selected from among the sum total of individuals of that generation." The useful variation is transmitted to the next generation and so on, until it becomes so marked as to give the organism possessing it the right to claim specific rank. You will observe that the Darwinian theory involves two assumptions, first, that any useful variation is inherited, and, second, that the variation once started goes on increasing and becomes more and more emphasised from generation to generation.

But according to the Mendelian hypothesis "heritable variation has a definite basis in the gamete, and it is to the gamete, therefore, not to the individual, that we must look for the initiation of the process. Somewhere or other in the course of their production is added or removed the factor upon whose removal or addition the new variation owes its existence. The new variation springs into being by a sudden step, not by a process of gradual and almost imperceptible augmentation. It is not continuous but discontinuous, because it is based upon the presence or absence of some definite factor or factors—upon discontinuity in the gametes from which it sprang. Once found, its continued existence is subject to the arbitrament of natural selection. If of value in the struggle for existence, natural selection will decide that those who possess it shall have a better chance of survival and of leaving offspring than those who do not possess it. If it is harmful to the individual, natural selection will soon bring about its elimination. But if the new variation is neither harmful nor useful there seems no reason why it should not persist. On the old view, no new character could be developed except by the piling up of minute variations through the action of natural selection. Consequently, any new characters found in animals or plants must be supposed to be of some definite use to the individual, otherwise it could not have developed through the action of natural selection. But there are plenty of characters to which it is exceedingly difficult to ascribe any utility, and the ingenuity of the supporters of this view has often been severely taxed to account for their existence. On the more modern view this difficulty is avoided. The origin of a new variation is independent of natural selection, and provided that it is not directly harmful, there is no reason why it should not persist. In this way we are released from the burden of discovering a utilitarian motive behind all the multitudinous characters of living organisms. For we now recognise that the function of natural selection is selection and not creation. It has nothing to do with the formation of the new variation. It merely decides whether it is to survive or to be eliminated" (Punnett).

You will now, I hope, have grasped how fundamental has been the change in our conception of the evolutionary process in consequence of the acceptance of the Mendelian principles. Time will not permit me to go into further detail on the subject, but I would remind you that on these principles rest the new sciences of Genetics and Eugenics, and recommend you to study Punnett's little work on Mendelism for yourselves. The books and papers quoted therein will guide you in your further investigations into this most interesting and important department of biological science.

LECTURE XI

MODERN VIEWS ON THE COMPOSITION OF CHLOROPHYLL

IN a previous lecture I attempted to trace very briefly the progress in our knowledge of chlorophyll and its functions down to the beginning of the twentieth century. I must now endeavour to tell you, even more briefly, the main additions made during the last two decades. I told you that, although quite a small army of chemists and botanists had been struggling to determine what chlorophyll really consisted of—was it one pigment or a mixture of several, what were its decomposition products, and so on—the problem was left in a condition so confused that the average botanical student, unskilled in the intricacies of chemical manipulation, was in a fair way to give up the whole subject as hopelessly incomprehensible. Of recent years, however, chiefly due to the labours of Willstätter and his pupils, we have at last apparently reached some measure of finality in this most difficult line of research, and the results achieved have, fortunately, been summarised for us by Willstätter himself, in conjunction with his colleague, Stoll, in their volume entitled *Untersuchungen über Chlorophyll, Methoden und Ergebnisse*, published in 1913. This work runs to over 400 pages, and therefore you will not expect me in a few minutes to give you more than the very briefest abstract of the authors' results.

The outstanding conclusions at which Willstätter arrives from an immense number of analyses is that leaf green is a complex pigment in which may be recognised four distinct components, viz. a-chlorophyll, b-chlorophyll, carotin, and xanthophyll. The chemical formulae for these pigments, as given by Willstätter, are: a-chlorophyll, $C_{55}H_{72}O_5N_4Mg$, blue green in tint; b-chlorophyll, $C_{55}H_{70}O_6N_4Mg$, pure green or yellow green; carotin, $C_{40}H_{56}$, orange red; xanthophyll, $C_{40}H_{56}O_2$, yellow. This was the composition of leaf green found in every case submitted to analysis. The only metal present was magnesium, and neither iron nor phosphorus could ever be detected, thus ending the long-drawn-out controversy as to the presence or absence of these elements that had taken place during the latter half of the nineteenth century. Willstätter also found that in 100 parts of dry leaf substance about 0·6 was composed of a-chlorophyll, about 0·2 of b-chlorophyll, about 0·07 to 0·12 of xanthophyll, and about 0·03 to 0·08 of carotin, and that they were invariably accompanied by sundry fatty or waxy impurities which had to be separated off before research could be carried out on the pigments themselves. Willstätter came to the conclusion also that chlorophyll occurs in the leaf in the colloidal condition, for the position of the bands in the absorption spectrum of fresh leaf green coincided exactly with the position of those of a colloidal solution of the pure a-chlorophyll.

Willstätter also examined the pigments of Algae and compared them with those extracted from Phanerogams. He found the same four pigments in Chlorophyceae, although the proportional amounts differed, while in Phaeophyceae he demonstrated the presence of a-chlorophyll, traces of b-chlorophyll, and also of carotin and xanthophyll, but established the existence of a third yellow pigment, fucoxanthin, which had a formula, $C_{40}H_{56}O_6$, while all the yellow pigments were relatively much more abundant than in Phanerogams.

The influence of acids and alkalies on the green pigment had occupied the attention of many workers during the preceding years, and the decomposition products had been studied in great detail, but the whole subject was in a hopeless state of confusion, until Willstätter, with the aid of ingeniously devised methods and excessive care in manipulation, elucidated the relationships of these manufactured products. I cannot do more here than summarise a few of his chief results in a sentence or two; for details I must refer you to the original volume, or to the excellent outline given by Jörgensen and Stiles in the *New Phytologist* for 1917.

When acted on by alkalies chlorophyll yields salts of acids known as "chlorophyllins" which contain magnesium. On heating these chlorophyllins with strong alcoholic alkalies a series of "phyllins," also containing magnesium, are obtained, and the magnesium is separated from them when they are treated with acids, the resulting compounds being "porphyrins."

When chlorophyll is acted on by acids, on the other hand, a body known as "phaeophytin" is produced, which possesses no magnesium, from which in turn by action of alkalies, among other derivatives, an alcohol called "phytol" is obtained. Further, Willstätter postulates an enzyme present in alcoholic solution which acts on the chlorophyll, producing a series of bodies known as "chlorophyllides," which were mistaken for crystalline chlorophyll by the earlier workers. The acid phaeophytin from a-chlorophyll forms olive green phytochlorin, and that from b-chlorophyll red phytorhodin.

Willstätter next proceeds to discuss the chemical and physical characters of all these derivatives, as also of the yellow pigments carotin and xanthophyll, which latter are stable in alkalies but unstable in acids. He also treats in great detail of the methods he and his fellow-workers adopted in isolating and purifying these various pigments and their derivatives. It is interesting to note that he emphasises the advantages of nettle leaves as sources of the pigments for laboratory investigation, for these leaves were also recommended by Stokes, sixty years previously, who also obtained two green and two yellow pigments from leaf green, as I have already told you.

A long series of experiments were undertaken to determine the variation in amount of the different constituents of leaf green, the results of which I have briefly summarised above. The time of the day at which the leaves are gathered does not seem to make any difference in the relative amounts of the pigments nor in the total quantities present. The relative proportion of the two chlorophylls also appears to be approximately constant both in different plants and in sun and shade leaves, but in the case of the yellow pigments the ratio between them seems to be greater in shade leaves, as is also the case in regard to the ratio between the total green and the total yellow pigments. Such quantitative determinations are of some importance seeing that earlier writers, such as Haberlandt, held that the intensity of photosynthesis varied with the amount of chlorophyll present in the leaf. Willstätter has carried out preliminary experiments on this subject, but as these investigations are, so far as I am aware, still incomplete, I do not propose to trouble you with them, beyond saying that it would appear that the amount of chlorophyll present does condition the amount of carbon dioxide absorbed, provided no other "limiting factor" intervenes to disturb the relationship.

A good deal of emphasis has been laid by many writers on the close similarity in chemical composition between the red colouring matter of the blood, haemoglobin, and chlorophyll, but, according to Professor Bayliss, Willstätter himself "does not regard the similarity . . . as being of any great significance."

Finally, recent research appears to show that the pigment is not diffused through the plastids, as was once believed to be the case, but forms "a highly concentrated layer on the surface of these bodies," and that the plastid itself "is practically solid."

The Functions of Chlorophyll

As you are well aware, in consequence of the activities of chlorophyll in sunlight, certain products result, oxygen gas is given off from the leaf and carbohydrates appear in it. Following on Priestley's historical discovery and Ingen-Housz's experimental investigations, many writers have dealt with the amount of oxygen exhaled as compared with the amount of carbon dioxide taken in, such as, in early days, De Saussure and, in later years, Bonnier and Mangin. The primary difficulty which confronted all the early workers was the separation of the gaseous interchanges connected with photosynthesis from those concerned with respiration. The first who made any successful attempt at disentangling the two processes were Bonnier and Mangin, and you will remember that the net result of their work was to show that the fraction $\frac{CO_2}{O_2}$ in respiration is less than unity, while the fraction $\frac{O_2}{CO_2}$ in photosynthesis is greater than unity, and that the real photosynthetic coefficient can be obtained only by a comparison of these two fractions. Bonnier and Mangin's work has quite recently been criticised by Maquenne and De Moussey. These authors hold that the real photosynthetic coefficient is approximately unity, but, despite the large number of cases examined, it cannot be said that the exact volume relationships of the two gases have even yet been determined.

As regards the carbohydrates formed, Sachs in 1862 held that starch was the first visible product of photosynthesis, but Meyer, in 1885, showed that in many plants starch is never formed at all and that such plants produce sugar instead. Ten years later, as the result of Brown and Morris's work, it had come to be believed that several sugars occurred in leaves, and the relationships of these to each other, to starch and to the activities of chlorophyll, were problems that greatly exercised the minds of chemists. Among the more recent contributions to the subject are the papers by Parkin (1911), who claims the presence of glucose and fructose in the leaves of the snowdrop, and of Davis, Daish, and Sawyer of Rothamsted, who doubt the presence of maltose, as asserted by Brown and Morris, but attempt to show that pentoses occur, though Brown and Morris failed to identify these sugars. With such contradictory statements before us it is obvious that the last word on the precise carbohydrates present in the photosynthetically active leaf has yet to be spoken.

Although Sachs had stated that starch was "the first visible product" of photosynthesis, he also expressly said that this did not exclude the formation of intermediate bodies between that very complex product and the primary water and carbon dioxide. That sugars were to be included among such transitional substances was apparent after Meyer had shown that some plants did not form starch at all, and that some which did could form it from a 10 per cent solution of fructose. Meyer, Schimper, and others in the later years of the nineteenth century proved conclusively that starch could be formed from sugar both by chloroplasts and by colourless amyloplasts, and this strengthened the view, originally suggested by Baeyer, that the primary product of photosynthesis might be formaldehyde, whence a hexose might arise by polymerisation, and which in turn might give origin to sucrose or to starch. In 1907 Strakosch, during his investigation of the sugars of the leaf, concluded that glucose was the first sugar to appear and that sucrose was an after-product. The reliability of the microchemical tests employed by Strakosch is, however, called in question by Davis and his co-workers, and by Mangham in 1915, for they point out that microchemical tests are of little value in determining the presence of one sugar when others are present also. To sum up, in the words of Jörgensen and Stiles, " While we may regard starch as a secondary product of assimilation, and while also there is good evidence that carbohydrates are translocated, as hexose sugars in some cases, or to some extent at any rate, and while there is strong evidence that sugars are the first definitely known products of the assimilatory process, there is no evidence at present as to which particular sugar is the first one to be produced in the leaf."

Energy and Photosynthesis

During the entire period that a leaf is exposed to light, radiant energy is falling on its surface ; part of that energy is transmitted through the leaf and part is absorbed. What becomes of the energy absorbed, and how can its amount be estimated ? Kreusler, in the later years of the nineteenth century, calculated the intensity of photosynthesis by measuring the amount of carbon dioxide absorbed by the leaf. In 1905 Brown and Escombe also employed this method, comparing the results they obtained with the observed increase in dry weight. Finding that these results did not correspond, they confined themselves to the first of the two methods and calculated the dry weight from the observed CO_2 intake, the increase in dry weight being taken as proportional to the amount of carbohydrate formed. Some doubt was thrown on the reliability of Brown and Escombe's results by Thoday in 1909, who found considerable variations in the proportions obtaining between the increase in dry weight and increase in carbon content in the leaf.

Several investigators (notably Puriewitsch, in 1914) have in recent years attempted to estimate the heat of combustion of photosynthetic products, with the object of gaining some knowledge of the energy relations of the leaf. Their results, however, do not support Brown and Escombe's figures, for the latter assumed that the heat of combustion of all the photosynthetic products might be comparable with that of glucose, which has been shown not to be the case.

An attempt to measure quantitatively the radiant energy falling on a unit area of leaf surface in a unit of time was made by Brown and Escombe in 1905, and to determine what proportion of this energy was employed in photosynthesis, how much in transpiration, and how much was otherwise disposed of. That used up in photosynthesis was, as we have seen, estimated by measuring the quantity of carbon dioxide absorbed and calculating the heat of combustion of the photosynthetic products formed. The energy used in transpiration was obtained by determining the amount of water transpired, and the heat necessary to bring about the vaporisation of that amount of water at a given temperature. These two totals subtracted from the total incident energy gave the amount lost in transmission, reflection, etc. Without going into the details of Brown and Escombe's experiments it may be said that they found that only a surprisingly small proportion of the radiant energy was used in photosynthesis, in one case only 0·6 per cent. Puriewitsch obtained considerably higher values in some cases, but neither Brown and Escombe nor Puriewitsch appear to have taken into account the influence of limiting factors, and thus their results cannot be regarded as conclusive.

Various attempts have also been made to estimate the energy value of the different wave-lengths of the spectrum. I have already stated to you the views of earlier workers like Daubeny, Draper, Sachs, Pfeffer, and Lommel as to the rays most efficacious in photosynthesis, and also the more detailed investigations carried out by Reinke, Timiriazeff, and Engelmann, who agreed in regarding the red rays absorbed by chlorophyll, viz. those between B and C, as the most important, though Engelmann claimed there was a second photosynthetic maximum in the blue violet region. The subject was re-investigated by Kniep and Minder in 1909, and from their experiments they conclude that, provided the light is of the same intensity, red and blue light produce the same photosynthetic results, but as they used the old gas-bubble method for estimating the intensity of photosynthesis, their conclusions cannot be said to be beyond criticism. You will see, therefore, that the new century has not as yet provided us with data on the energy relations of the leaf at all comparable with those of Willstätter on the chemical side, and it is more than strange that no one with the requisite physical knowledge seems to have considered it worth while to determine whether any of the radiant energy is transformed into electric energy—taking into account at the same time the old work of Brodie and the newer experiments of Loeb on the decomposition of carbon dioxide and water by silent electric discharge. I drew attention to this point in 1908, but, so far as I know, the matter has been left untouched, and various circumstances have prevented me from following up the subject.

I am not going to trouble you with the speculations of Van't Hoff, Siegfried, and Willstätter on the mode by which photosynthesis is carried out, for, in the words of a recent writer, "It can safely be said at the outset that, when critically considered from a physiological view point, none of the existing theories is even moderately well established by observations of facts." In the Croonian lecture to the Royal Society in 1903 Timiriazeff made a witty reference to the philosophers of Balnibarbi, whom Gulliver in his travels found instructing their pupils how to extract sunbeams from green cucumbers, and suggested that we also must rest content to contemplate for a little while longer green leaves locked up in glass bottles, ere we reached a clear interpretation of the mysterious performances that take place in them when they are exposed to sunlight.

Sensitivity and Movement

We have seen that before the end of the nineteenth century it had come to be generally accepted that all organisms, whether plant or animal, were sensitive to external stimuli and responded in various ways, visible or invisible. It was also recognised that it was the protoplasm that possessed the capacity for appreciating or perceiving the stimulus, but although this power was obviously associated in animals with certain differentiated sense organs, the existence of similar structures in plants was more difficult to establish. It is only within recent years that special sensory cells or tissues have been postulated for plants. Then again the sensory organs in animals were well known to be continuous with definite tracts of afferent nerve tissue which transmitted the excitations induced by the stimuli to nerve centres, and thence by a system of motor or efferent nerves to the seats of response, *i.e.* of movement, secretion, or what not. Nerves and nerve centres in the plant seemed, however, to be conspicuous by their absence. Finally, although plants had undoubtedly the power of movement, to some limited extent at least, the special contractile tissues so apparent in the animal could not be identified in the motile organs of plants.

So far as sense organs are concerned we are justified in saying that, as a result of the investigations of Pfeffer, Haberlandt, Borzi, and others, certain sensitive organs, such as tendrils, possess what Haberlandt calls "tactile pits," "tactile papillae," hairs or bristles or "stimulators." These latter organs are specially noticeable in carnivorous plants, such as Dionaea, and in various genera of orchids, where they are associated with intricate movements connected with cross pollination.

There are two stimuli in nature to which plants are peculiarly sensitive, viz. gravity and light. The typical root is positively geotropic and the shoot axis negatively so, while dorsiventral leaves are diageotropic. These fundamentals were made clear to us by Knight early in the nineteenth century, and the discovery of the sensitive protoplasm by Von Mohl enabled physiologists to recognise that "vehicle of irritation" of whose existence Knight could not convince himself. The next question that arose was, what was it that stimulated the protoplasm? An answer to this question was offered in 1900 by Noll, Haberlandt, and Němec. These investigators held that each organ that responded to the stimulus of gravity possessed certain sensory cells or "statocysts," the external layer of whose protoplasm was peculiarly sensitive to the impacts of minute granules, such as starch grains, to which the name "statoliths" was given. The theory is expressed by Haberlandt in the following words:

"When an organ containing statocysts is in a condition of geotropic equilibrium, the pressure of the starch grains against the physically lower portion of the ectoplast remains unperceived, or at any rate leads to no responsive movement. But as soon as the part under consideration is displaced from its stable position, the starch grains fall against that portion of the ectoplast which is now on the physically lower side of the cell; a new and unfamiliar state of stimulation is thereby produced, with the result that a geotropic movement takes place, which brings the organ back into its former state of equilibrium."

The transference of the starch grains from the original to the new position occupies some time, and, as they continue falling on the new sensitive area, the stimulus increases in intensity. This period, which lasts from five to twenty minutes according to Haberlandt, is called the "presentation time," that is, the period of time during which the stimulus must be applied before any visible movement ensues. This so-called "statolith theory" is a very alluring one, and has attracted the attention of many investigators since it was first put forward, some supporting, some attacking the theory itself, as well as the facts on which it is supposed to rest.

Knight, you will remember, was also greatly exercised over the perverse behaviour of vine leaves, that would insist on attempting to defy his efforts to make them grow at other angles with incident light than those to which they were accustomed in nature. Here again the discovery of protoplasm presented the "vehicle of irritation," but later on the further question arose of the existence or non-existence of specific sense organs on such leaves as responded to the phototropic stimulus. Once more Haberlandt came forward, in 1905, with a theory to explain the mode of stimulation of the protoplasm. According to him the epidermal cells, of many leaves at least, possess papillate elevations which he thought acted as lenses, concentrating the incident rays on the sensitive ectoplast. In the phototropic rest position the light is focussed on the centre of the ectoplast, and in that position produces no response; but when the angle of the sensitive organ to the light ray is changed the beam of light strikes the ectoplast at some other spot, and induces an excitation, and finally an endeavour on the part of the organ to regain the original orientation. This also is a most interesting line of investigation and worthy of your study, but it would be quite out of place for me to discuss it here.

When the terminal leaflet of a "Sensitive Plant" is gently touched with a slightly heated wire the successive pairs of leaflets fold together, and when all have so folded the entire leaf droops. The movement may, if the stimulus be sufficiently intense, extend to other leaves also. It follows, first, that the excitement set up by the stimulus—the "excitation"—must have been conveyed or transmitted from the original point of application to all these regions of response, and, second, that there must exist some mechanism by which the movements are carried out; in other words, the plant must possess something analogous at least to nerves and contractile tissue. Whatever the nature of the transmitting agency may be, we have to admit that it is nothing like so efficient as the corresponding agency in an animal. In the human nerve the rate of transmission has been estimated at about 50 metres (160 feet) per second, while that in the "Sensitive Plant" is variously estimated at from 1·5 cm. to 10 cm., *i.e.* at most 4 inches per second. You will remember that Tangl, Gardiner, and others long ago demonstrated the existence of exceedingly delicate protoplasmic threads connecting the protoplasts of adjacent cells, and many authorities, notably Strasburger, Czapek, Hill, and Haberlandt, consider that the excitation may be transmitted along such intercellular protoplasmic fibrillae. Němec, in the beginning of the present century, went a step further and claimed to have demonstrated very fine fibrillae or longitudinal protoplasmic tracts in many elongated cells, extending from one transverse wall to the opposite one, but Haberlandt failed to find conductors in the sensitive organs he examined, with the solitary exception of Mimosa, the "Sensitive Plant." In this much-investigated plant Haberlandt considers the transmitting elements to be certain elongated tubular cells with delicate pitted walls associated with the bast region of the vascular bundles. Haberlandt explains the transmission of the excitation in the following words:

"When the pulvinus of a pinnule moves upwards in response to a shock, pressure is exerted upon the highly turgescent transmitting cells, partly owing to changes in the form and volume of the relaxed half of the pulvinus, and partly owing to the mechanical effects of the curvature; the local rise of pressure produced in this way will be propagated along the system of tubes, owing to the elastic tension of their walls, in the form of a pressure wave, like the pulsations which travel through the arterial system of an animal. This wave of compression or positive tension acts like a shock stimulus upon the nearest pulvinus, and so leads to an indirectly induced responsive movement. The initial rise of pressure which starts the wave is not large; hence the stimulus, while extending from one pair of pinnules to the rest, does not convey a perceptible shock to the comparatively insensitive pulvinus of the subpetiole, and never penetrates as far as the main pulvinus. Traumatic stimulation—such as may be caused by the severance of a pinnule—instantly destroys the turgor of the injured transmitting cells; as a consequence a large local fall of turgor results, which is propagated through the tubular system as a wave of relaxation or negative tension. The initial change of pressure caused by such a mechanical injury is comparatively large; hence a much more violent disturbance is produced in the adjacent pulvini than can possibly arise when the primary stimulus is due to shock. A traumatic stimulus can accordingly be transmitted over a relatively large distance; it will not only reach the main pulvinus, but may also travel through the stem to other leaves."

Haberlandt claims that the mode of transmission is hydrodynamic, and states that when the transmitting tubes are cut a drop of transparent liquid escapes which is not derived from the xylem elements, and which is chemically cell sap, not water. In a paper published in 1916 Ricca, an Italian botanist, criticises the whole of Haberlandt's work and contradicts his statement that the fluid does not escape from the xylem. Ricca obtained responses quite as rapid with decorticated petioles as with corticated ones, and holds that the excitation is transmitted by the xylem and not by Haberlandt's leptome tubes. By means of ingeniously devised experiments, Ricca also shows that after a length of tissue had been exposed to a temperature of 150° for thirty hours the leaflets closed at night and reopened next day, so that the stimulus must be transmitted by non-living elements, and intercellular protoplasmic fibrillae are excluded. Macdougal also showed that stimuli were transmitted through tissues killed by heat. Ricca's view is that there are certain substances in the cell sap which he considers homologous with the animal "hormones," and which, after they have entered the xylem vessels, may be transmitted to and stimulate the motile organ. Boysen, Jensen, and De Paál have also suggested that certain tropisms may be explained by the transmission of specific substances. If this be so we must revise completely all our ideas as to stimulus and response in plants, but the data available do not warrant us as yet in coming to any general conclusions on the subject.

Ecology

In 1895 Warming of Copenhagen published a book which he termed *Plantesamfund* or *Plant Communities*, which presented for the first time a detailed exposition of what has come to be known as "Ecology"—a subject which discusses "how plants or plant communities adjust their forms and modes of behaviour to actually operating factors, such as the amounts of available water, heat, light, nutriment, and so forth." Warming's pioneer work made its appearance, in a much improved and extended form and in English dress, in 1909. In 1898, soon after Warming's original treatise was published, another work of great importance appeared, *Plant Geography, upon a Physiological Basis*, from the pen of the talented young botanist A. F. W. Schimper, who died in 1901 at the early age of 45. I think it may be safely said that to these two works we owe the foundations of our knowledge of ecology, a subject which bulks so largely in modern educational curricula, where botany figures as a component. What this new sub-science deals with is indicated not only by the definition I have just quoted to you from Warming's work, but also by the practical knowledge you have acquired in the field. In no country, save perhaps America, has this subject been more assiduously studied than in Britain, a result due in large measure to the efforts of Tansley and Oliver. How much has already been accomplished in this relation in the British Isles, and, incidentally, how much still remains to be done, you may learn for yourselves from the volume edited and in part written by the first named of these workers, in 1911, entitled *Types of British Vegetation*. The task before the genuine ecologist is by no means an easy one, for, as Tansley puts it, "The chief obstacle to the rapid development of ecology on fundamental lines is the laborious and time-consuming nature of the work and the chemical and physical training required for its prosecution." By no means all those who have undertaken research in this direction have sufficiently appreciated this warning, and, as a consequence, the periodic journals not infrequently contain pages of observations purporting to be ecological investigations which are without any permanent value. One cannot help regretting the fact also that some ecologists deem it necessary to employ a terminology so uncouth as to discourage the would-be student, and give him a distaste for the subject from the very outset. *Research Methods in Ecology*, published in 1905 by Clements, is a case in point.

LECTURE XII

THE PTERIDOSPERMS AND THE SEED

"THERE is no part of Fossil Botany in which there have been such revolutionary changes within a very short period as in the question of the position of Palaeozoic Ferns. Till within the last three years [as from 1906] the Ferns were universally regarded as forming one of the dominant classes of Palaeozoic plants—in fact, the most dominant of all—and this estimate of their importance will be found in all the Text-books. According to the computations of systematists the Ferns constituted almost exactly one-half of the known Carboniferous flora. The position has now so completely changed that Professor Zeiller, than whom there is no higher authority, wrote, in August of last year [1905], that the Ferns of the Palaeozoic period, though 'they were probably not entirely absent, occupied an altogether subordinate rank.' The ground for the radical change of view which Professor Zeiller's words indicate, is, of course, to be found in the recognition of the Pteridosperms, a class of seed-bearing plants, to which, as it now appears, the great majority of the supposed Palaeozoic Ferns belonged."

In these words Dr. Scott introduces to us a sketch of the remarkable discoveries that have been made of late years relating to the origin of seed-bearing plants. If we are to look for the point of origin of the Spermatophyta away deep down in the Palaeozoic strata, then our notions as to the antiquity of that phylum must be radically altered, and for that reason the discoveries I refer to are among the most important ever made in plant morphology.

In Lecture VIII. I mentioned the name of Binney, a Lancashire lawyer, who had been attracted to the study of fossil trees in the St. Helens Coalfields, as the discoverer (in 1866) of a petrified stem which received the name of *Lyginodendron Oldhamium.* This stem was described by Williamson a few years later, and in greater detail by Williamson and Scott in 1895, when it was shown to possess anatomical characters which connected it with the Cycadaceae, with hints, in the young shoot, of the Osmundaceae. The first step in placing this remarkable plant in its proper category was the recognition of the fact that its foliage was that of a "fern" long known as *Sphenopteris Höninghausi*, while the superficial prominences on both stem and leaves led to the identification of the "fern" petiole, known as *Rachiopteris aspera*, as the leaf-stalk of Sphenopteris, and hence of Lyginodendron. The plant was thus seen to combine both Cycadaceous and fern-like vegetative characters, and to such synthetic forms Potonie, in 1897, applied the group name Cycadofilices.

Some of these epidermal outgrowths were discovered to be of the nature of stalked glands, and glandular structures of the same character were shown by Oliver and Scott, in 1903, to be developed on the investing cupule of a seed which Williamson had found and named *Lagenostoma Lomaxi.* Oliver and Scott, from the occurrence of these unique glands, as also from a detailed investigation of the anatomy of the vascular system of the peduncle, had no difficulty in proving that the seed Lagenostoma was the fructification of Lyginodendron. As the seed is now described and figured in every botanical textbook I need not do more than say that it presents all the fundamental features of a Cycad seed, pollen chamber, central nucellus column and all. Here, then, was a fossil tree with pronounced secondary thickening, with the petioles and leaves of a fern and the seed of a Spermatophyte. What, then, were the male organs or stamens like?

Among the ancient fern-like fronds that were still regarded as remains of true ferns were types representative of our modern Marattiaceae. Among these was a genus, Crossotheca, whose pinnules bore half-a-dozen or more pointed microsporangia. The Scotch palaeobotanist, Kidston, made a detailed examination of this frond in 1905, and proved that it was not a Marattia but the male frond of Lyginodendron, so that, while the megasporangium had adopted seed characteristics, the microsporangium had remained fern-like.

The labours of palaeobotanists during recent years have considerably extended our knowledge of this ancient family, now known as Pteridosperms, and several other series of stems, leaves, and seeds have been correlated and pieced together, but for details as to these I must refer you to Scott's *Fossil Botany* wherein you will find the evidence given in full. As to the affinities of the Pteridosperms I cannot do better than quote some sentences from Scott's article in the *Progressus rei botanicae*, from which, indeed, I have already read you an extract.

"A more fundamental question is that of the relation of the Pteridospermeae to the Cryptogams. As has been sufficiently shown in the preceding pages, all the characters in which the Pteridosperms show Cryptogamic affinities, whether in anatomical structure, in the morphology of the sporophyll, or in the nature of the male fructification, point clearly to their derivation from ancestors belonging to a Filicinean stock. They have been described as 'Ferns which have become Spermophytes,' and the phrase is appropriate. When, however, we come to inquire into the characters of the Filicinean group from which the Pteridosperms arose, we find that our data are insufficient. They are themselves, in all probability, as ancient as any land plants known to us, and their actual origin lies further back than our records at present extend. Considering that some of the Pteridosperms show a decidedly simple anatomical structure (as in Heterangium) we may assume that they were derived from plants of a simple type of organisation. It would be rash in the extreme to identify any of the known 'Primofilices' with the ancestors of the Pteridosperms; they are not nearly old enough geologically, and our knowledge is much too narrow to enable us to determine how far they may have retained the characters of the original common stock. The utmost we can venture to say is, that these simpler Palaeozoic Ferns, the Botryopterideae and their allies, probably stand nearer the Cryptogamic progenitors of the Seed-plants than any other group of which the record has come down to us.

"Where we find among the Pteridosperms characters resembling those of more advanced Filicinean types, they are probably to be attributed to parallel development rather than to inheritance. The 'polystely' of Medullosa, for example, if, as there is reason to believe, it arose within the Pteridospermic family Medulloseae, was not a directly inherited Filicinean character, but rather a new development on Filicinean lines.

"We may sum up the position of the question as to the derivation of the Pteridosperms in the statement that all the evidence points to their having sprung from the same stock with the Ferns. The antiquity of the Ferns, and especially of the comparatively simple types represented by the Botryopterideae and related forms, appears sufficiently established to afford an historical basis for this conclusion."

Bennettites and the Primitive Flower

While progress had thus been made not only in tracing back the origin of the seed habit but also in reconstructing the Flora of the Carboniferous epoch, certain other discoveries began to throw light on the much-debated question as to the origin of the Angiosperms.

In the Mesozoic rocks remains of many Cycad-like plants had been found whose fructifications, however, did not seem to bear any resemblance to those of recent forms. One of these fossils was described by Carruthers as far back as 1868 under the name of Bennettites, known at a later date also under the name Cycadeoidea, more especially in America, where plentiful remains had been discovered. So far as its stem structure was concerned it seemed to fit into the Cycad series, although its anatomy was of a somewhat simpler type; the chief point of difference seemed to be that it possessed many lateral branches whose origins were imbedded between the leaf bases with which the surface of the stem was covered. These lateral branches carried the fructifications which were conical or pyriform structures, consisting of a convex thalamus whence arose externally a number of bracts enclosing a large number of seeds with long funicles, separated from each other by interseminal scales whose distal ends expanded, closely covering the terminal seeds, and forming a sort of pseudo-pericarp through which the micropyles of the seeds protruded. As in the case of the Pteridosperms, I must refer you for details to the published accounts of this remarkable plant, otherwise these lectures will tend to become a textbook of botany, to write which is not at all my intention.

The seed contained a well-marked dicotyledonous embryo with scarcely any endosperm, a fact first recognised by Solms Laubach in 1890; in this respect it differs from all members of the modern Gymnosperm series.

In 1899, owing to the researches of Wieland, we began to learn something of these Cycad-like plants as they were represented in the Mesozoic strata of Dakota. In that year Wieland discovered the stamens, and in

1901 he found these associated in the same specimen with the previously known gynaecium, so that we were presented with a Cycad-like plant bearing a hermaphrodite flower! The stamens were twenty or more in number, arranged in a single whorl arising from the base of the seminal cone, their bases being coherent into a cup. Each stamen was a pinnate leaf bearing two rows of synangia on each pinna, and the whole flower was enclosed in several series of hairy perianth segments. It is only natural that this remarkable flower should be interpreted in different ways by different authorities. All are agreed that the pinnate microsporophylls correspond to stamens, but some, *e.g.* Lignier, consider the interseminal scales as modified leaves, and each stalked seed as a megasporophyll derived from a bud with only one leaf, arising in the axil of the interseminal scale. Others, however, think that all the organs springing from the thalamus are foliar, and that the entire structure corresponds to one strobilus or flower. This latter view, which certainly seems the most reasonable, is held by Wieland, Scott, and by Arber and Parkin. Indeed the last-mentioned authors in a paper on *The Origin of the Angiosperms*, published in 1907, term the flower of Bennettites a "proanthostrobilus," leading up to a hypothetical anthostrobilus and the modern flower, such as we meet with in the Magnoliaceae.

As to the general relationships of these primitive "flower-bearing" Cycads with the modern Angiosperms I cannot do better than quote to you Scott's summary as given in the concluding chapter of his valuable *Studies in Fossil Botany.*

"Taking the whole of the characters into consideration, the evidence of affinity between the Mesozoic Cycadophyta and the Angiosperms appears very strong. It cannot, of course, be supposed that the Bennettiteae were on the direct line of Angiospermous descent, for there are manifest points of difference, notably in the great complexity of the stamens and in the organisation of the ovary-wall or pericarp, which was not formed by the carpels themselves, but by the associated sterile scales. There may be a difference of opinion as to the nearness of the relation between Bennettiteae and the higher 'Flowering Plants,' but the points of agreement are so striking that we can hardly fail to recognise that a real relation exists, and that the ancestry of the Angiosperms, hitherto the most obscure subject in the phylogeny of plants, is to be sought among the great plexus of Cycadophytes, which overspread the world in the Mesozoic Period.

"This conclusion opens up the question of the relation of Monocotyledons to Dicotyledons. If the Angiosperms were derived from Cycadophyta, it would appear to follow that the Dicotyledons were first evolved, for their structure has clearly much more in common with the Cycad type than that of the Monocotyledons. The latter would thus be regarded as a branch line of descent, diverging, no doubt at a very early stage, from the main Dicotyledonous stock. This view has been maintained, on other grounds, by various modern botanists. So far, however, as the palaeontological record shows, the two classes are of almost equal antiquity, both appearing for the first time in the Lower Cretaceous rocks. By the Upper Cretaceous age the Angiosperms had already seized the dominant position which they now hold; the Monocotyledons were always subordinate in numbers to the other class, but the occurrence of typical Palm-wood in Cretaceous rocks is a striking proof of the early evolution of one of the most characteristic Monocotyledonous families.

"We are thus led to the conclusion that the whole of the dominant sub-kingdom of Flowering Plants, if akin, as we believe, to the Cycadophyta, belongs ultimately to the phylum which takes its name from the Ferns. We may add that the Gymnosperms, as a whole, may be referred to the same stock, for evidence has recently been adduced that the small group of the Gnetales (the only outstanding Gymnospermous family) may have been derived, by reduction of the floral organs, from forms allied to the Bennettitales.

" It thus appears, if the views here taken are justified, that the whole of the Spermophyta, whether Angiospermous or Gymnospermous, were ultimately derived, through primitive Seed-plants of the nature of Pteridosperms, from the same stock with the Ferns."

The Phylogeny of the Angiospermae

If, as you will now see, there is good reason to believe that the Bennettitales represent a very ancient type of seed plant which possessed a hermaphrodite flower, and which came off from a stock that gave origin to our existing flowering plants or Angiosperms, it follows that we must revise completely our ideas on the phylogeny and classification of these plants, and reject both the schemes that are at present in common use. The majority of English botanists have clung tenaciously to the system of Bentham and Hooker in some form or another. One of the most recently published text-books (that of Ganong, in 1917) emphasises the division of Angiosperms into " two sharply marked sub-classes," Monocotyledons and Dicotyledons, and goes on to treat of the former group first, starting with such "orders" as Pandanales, Naiadales, Graminales, Palmales, and Arales as " primitive " Monocotyledons, leading up to " differentiated " and " specialised " orders. The Dicotyledons are similarly treated and discussed under " primitive " orders, " differentiated " orders, and " specialised " orders. The " primitive " Dicotyledons are held to include such plants as the peppers, willows, beeches, alders, elms, figs, and a host of others with correspondingly low floral organisation. In this respect Ganong follows Engler's system which is so much in vogue on the Continent and in America. In our own country also Engler's system is making headway, and as recently as 1915 a *Synopsis of the Families of British Flowering Plants, based upon the System of Engler,* made its appearance, presumably for students' use.

If the word "primitive" has any meaning at all, and we are to regard peppers, willows, and the like as the earliest Dicotyledons and consequently the earliest Angiosperms (for Ganong admits that "the evidence suggests the derivation of the Monocotyledons from the Dicotyledons"), then the phylogeny suggested by a study of the Bennettitales is out of the question. Though I would warn you that I am expressing merely my own personal opinion, still I have no hesitation in prophesying that in years to come botanists will regard Engler's system as having done as much to retard the attainment of a true phylogenetic classification of Angiosperms as Linnaeus's sexual system retarded a natural classification, as it was then understood, in the eighteenth century. There are, however, a few, but very few, botanists that have not succumbed to the authority of Engler's great name in Taxonomy, and we owe a debt of gratitude among such to Lotsy, who has boldly accepted the newer doctrines, and based the classification of Angiosperms, in his great work *Botanische Stammesgeschichte,* on the hypothesis that Bennettites-like Dicotyledons represent the lowest and most primitive of the modern Angiosperms. The volume dealing with this most interesting and important question was published in 1911, but unfortunately the great European War has delayed the publication of the fourth volume of the treatise, and we still await Lotsy's views on the phylogeny of the highest groups of flowering plants. I do not think I can do better than sketch out for you very briefly the main lines of the argument for the view I have just stated, as given by Lotsy.

After giving an outline of Von Wettstein's fanciful

theory of the derivation of an Angiosperm flower from an inflorescence of unisexual flowers, Lotsy remarks that the author bases his view that the primitive Angiosperm flower was unisexual on the supposed origin of such Angiosperms from the Gymnosperms, where hermaphrodite flowers do not occur. The so-called "Monochlamydeae" are thus to be regarded as primitive, and amongst these the anemophilous forms stand lowest of all. The Monochlamydeae undoubtedly possess very simple flowers, but is simplicity, asks Lotsy, an invariable indication of primitiveness? May it not arise, secondarily, by reduction? May not anemophilous flowers be derived from entomophilous? Lotsy quotes at length, and with approbation, from a paper by Robertson, published in 1904, on "The Structure of the Flowers and the Mode of Pollination of the Primitive Angiosperms" in which the author attempts to demonstrate (and, as it appears to me, with success) that the primitive Angiosperms were entomophilous, and that the Monochlamydeae are not primitive but greatly reduced types.

The leading exponent of the newer view is undoubtedly Hallier, whose phylogeny Lotsy, in the main, follows. Hallier's views—which have been, strange to say, ignored by Engler—are as follows. I give you them in his own words quoted from an article on "The Natural (Phylogenetic) System of Flowering Plants" published in 1905.

"The Angiospermae are a natural (monophyletic) group, and not a polyphyletic one, as suggested by Engler in Engler and Prantl's *Natürlichen Pflanzenfamilien*. The Amentaceae are not to be considered as old types, remaining in a lower state of development, and allies or descendants of Gymnospermae, but, on the contrary, as the highest and most reduced types of one of the lines of Dicotyledons. They and all the other lines of Dicotyledons have been developed by reduction in flower and fruit from the Polycarpicae, the latter group being derived immediately from Bennettitaceae or other extinct Cycadales. In the same manner the Liliiflorae and all the other syncarpous Monocotyledons have been derived by union of the carpels, by reduction in the number of parts, by epigynous insertion of the perianth, and by other changes in the structure of the flower and fruit, from the polycarpous Monocotyledons (Helobiae), which latter group originated immediately from the polycarpous Dicotyledons (Polycarpicae and Ranales). In the Dicotyledons the Apetalae and Sympetalae are unnatural groups of polyphyletic origin."

As Hallier rightly points out in the article from which I have just quoted, "there can exist only one really natural system, namely, that which is identical with the tree of descent; to reconstruct this, systematic botany should be founded on a much broader and more universal base than at present, comprehending not only the morphology of the reproductive organs, but also all the other branches of botany, such as the comparative morphology of the vegetative organs; comparative anatomy, ontogeny and embryology; phyto-chemistry; physiology and ecology; structure of pollen and seed coat; relations to climate, seasons and to the surrounding organic world; plant geography; palaeophytology, etc." But this is the labour not of one man but of an army; the task not of one lifetime but of several generations of men. Then, but not till then, may we hope to see in its entirety a genealogical tree of the plant world spread out before us, with reasonable grounds for the belief that it represents, with some degree of accuracy, the succession and lines of descent of plant forms in past ages of the earth's history.

CONCLUSION

In the preface to a recently published and very excellent work on the History of Science by Prof. Libby, the following sentences occur:

"The history of science is an aid in scientific research. It places the student in the current of scientific thought, and gives him a clue to the purpose and necessity of the theories he is required to master. It presents science as the constant pursuit of truth rather than the formulation of truth long since revealed; it shows science as progressive rather than fixed; dynamic rather than static, a growth to which each may contribute. It does not paralyse the self-activity of youth by the record of an infallible past.

"It is only by teaching the sciences in their historical development that the schools can be true to the two principles of modern education, that the sciences should occupy the foremost place in the curriculum, and that the individual mind in its evolution should rehearse the history of civilisation. . . . The history of science studies the past for the sake of the future. It is a story of continuous progress. It is rich in biographical material. It shows the sciences in their inter-relations, and saves the student from narrowness and premature specialisation."

These sentences express exactly the conceptions that prompted the compilation of these lectures. They will have fulfilled their purpose if they have given you, even though in the briefest outline, some idea of the stages through which our knowledge of botany has passed, and be the means of tempting you not only to enquire more fully into the details of these successive phases, but also to resolve to add some fragment, however small, to the building of the temple of scientific truth.

BIBLIOGRAPHY

THE following brief list of books and memoirs is intended merely as a preliminary guide to the botanical student who may desire further details on the matters dealt with in each lecture. In these he will find abundant references to the original authorities. The selection of the present list has been carried out on three principles: (1) that the work shall be in English, (2) that it shall be readily accessible in any reasonably well-equipped botanical library, (3) that it shall be such as any senior undergraduate might be expected to consult. The individual works quoted, after being once indicated, are referred to in subsequent lectures by the prefixed number.

LECTURE I.

1. LEE GREENE, E. Landmarks of Botanical History. Washington, 1909.
2. ARBER, A. Herbals, their Origin and Evolution. Cambridge, 1912.
3. HORT, A. Theophrastus. Enquiry about Plants. London, 1916.
4. MIALL, L. C. The Early Naturalists, their Lives and Work. London, 1912.
5. OLIVER, F. W., and others. Makers of British Botany. Cambridge, 1913.
6. REYNOLDS GREEN, J. A History of Botany in the United Kingdom. London, 1914.
7. SACHS, J. History of Botany. Oxford, 1890.

LECTURE II., 1, 2, 4, 5, 6, 7.

8. GREW, N. The Anatomy of Plants begun. London, 1671.

LECTURE III., 4, 5, 6, 7.

9. HALES, S. Vegetable Staticks. London, 1731.
10. SCHLEIDEN, J. Principles of Scientific Botany. London, 1849.

LECTURE IV., 4, 5, 6, 7.

11. PRIESTLEY, J. Experiments and Observations on Different Kinds of Air. London, 1771.
12. HARVEY-GIBSON, R. J. Pioneer Investigators in Photosynthesis. New Phytologist, 1914.
13. INGEN-HOUSZ, J. Experiments upon Vegetables. London, 1779.

LECTURE V., 4, 5, 6, 7, 10.

14. SCOTT, D. H. Presidential Address. Proc. Linn. Soc., 1911.
15. VON MOHL, H. The Vegetable Cell. London, 1852.

LECTURE VI., 7, 14.

16. BOWER, F. O. The Origin of a Land Flora. London, 1908.
17. CAMPBELL, D. H. Mosses and Ferns. London, 1905.
18. HOFMEISTER, W. The Higher Cryptogamia. Ray Society, 1862.
19. REYNOLDS GREEN, J. History of Botany. Oxford, 1909.
20. SEWARD, A. C. Fossil Plants. Cambridge, 1897–1919.
21. WARD, L. Sketch of Palaeobotany. U.S. Geological Survey, Rept. V.
22. DARWIN, C. On the Origin of Species by means of Natural Selection. London, 1859.
23. DARWIN, F. The Life and Letters of Charles Darwin. London, 1887.
24. HUXLEY, T. H. Darwiniana. London, 1893.
25. WALLACE, A. R. Natural Selection and Tropical Nature. London, 1891.

LECTURE VII., 16, 17, 18, 19.

26. BOWER, F. O. Botany of the Living Plant. London, 1919.
27. COULTER, J. M., and CHAMBERLAIN, C. J. Morphology of Spermatophytes. New York, 1901.
28. LANG, W. H. An Ontogenetic Theory of Alternation of Generations. New Phytologist, 1909.
29. KNUTH, P. Handbook of Flower Pollination. Oxford, 1906.

LECTURE VIII., 6, 16, 17, 19, 20.

30. SCOTT, D. H. Studies in Fossil Botany. London, 1919.
31. DE BARY, A. Comparative Anatomy of Phanerogams and Ferns. Oxford, 1884.
32. HABERLANDT, G. Physiological Plant Anatomy. London, 1914.

33. JEFFREY, E. C. The Anatomy of Woody Plants. Chicago, 1917.
34. TANSLEY, A. G. The Evolution of the Filicinean Vascular System. New Phytologist, 1907–8.
35. PFEFFER, W. The Physiology of Plants. Oxford, 1900.
36. SACHS, J. Textbook of Botany. Oxford, 1874.

LECTURE IX., 7, 19, 26, 32, 35.

37. DIXON, H. H. Transpiration and the Ascent of Sap in Plants. London, 1914.
38. SACHS, J. Lectures on the Physiology of Plants. Oxford, 1882.
39. JÖRGENSEN, I., and STILES, W. Carbon Assimilation. New Phytologist, 1914.
40. BAYLEY BALFOUR, I. Pringsheim's Researches on Chlorophyll. Quart. Jour. Micr. Sc., 1880.

LECTURE X., 19.

41. REYNOLDS GREEN, J. The Soluble Ferments and Fermentation. Cambridge, 1899.
42. BATESON, W. Mendel's Principles of Heredity. Cambridge, 1909.
43. DE VRIES, H. The Mutation Theory. London, 1905.
44. PUNNETT, R. C. Mendelism. Cambridge, 1912.
45. WEISMANN, A. The Germ Plasm. London, 1893.

LECTURE XI., 19, 32, 35, 38, 39.

46. BAYLISS, W. M. Principles of General Physiology. London, 1915.
47. SCHIMPER, A. F. W. Plant Geography. Oxford, 1903.
48. WARMING, E. Oecology of Plants. Oxford, 1909.

LECTURE XII., 19, 20, 26, 27, 30.

49. HALLIER, H. The Natural (Phylogenetic) System of Flowering Plants. New Phytologist, 1905.
50. SCOTT, D. H. The Present Position of Palaeozoic Botany. Prog. Rei Botan., 1907.

EXPLANATION OF TABLE

THE appended table attempts to give a résumé of the development of the science of Botany in phylogenetic form. It must be clearly understood, however, that the names selected are merely representative, and are not to be taken as exclusive of others holding similar views and, possibly, of equal merit. When the same name occurs twice on the same phylum or on two different phyla the intention is to indicate that the author quoted was instrumental in advancing the science in more than one aspect. Thus, Brongniart's name is associated chiefly, perhaps, with important additions to our knowledge of fossil plants, but he also formulated a classification of the vegetable kingdom which must be taken into consideration in any outline of the development of the subject of taxonomy. Again, Bower's name appears twice on the "Cryptogamic" phylum; the earlier entry refers to his monographs on "Spore-producing members," the later to his "Origin of a Land Flora." Save in the very earliest entries, the dates are approximate only, and suggest the period rather than the exact year of publication of important contributions; in no case do they indicate the date of birth or of death of the author quoted. If more space were available the table might be extended, almost indefinitely, but for the ordinary student's purposes it is considered that the selection may suffice to provide landmarks in botanical history.

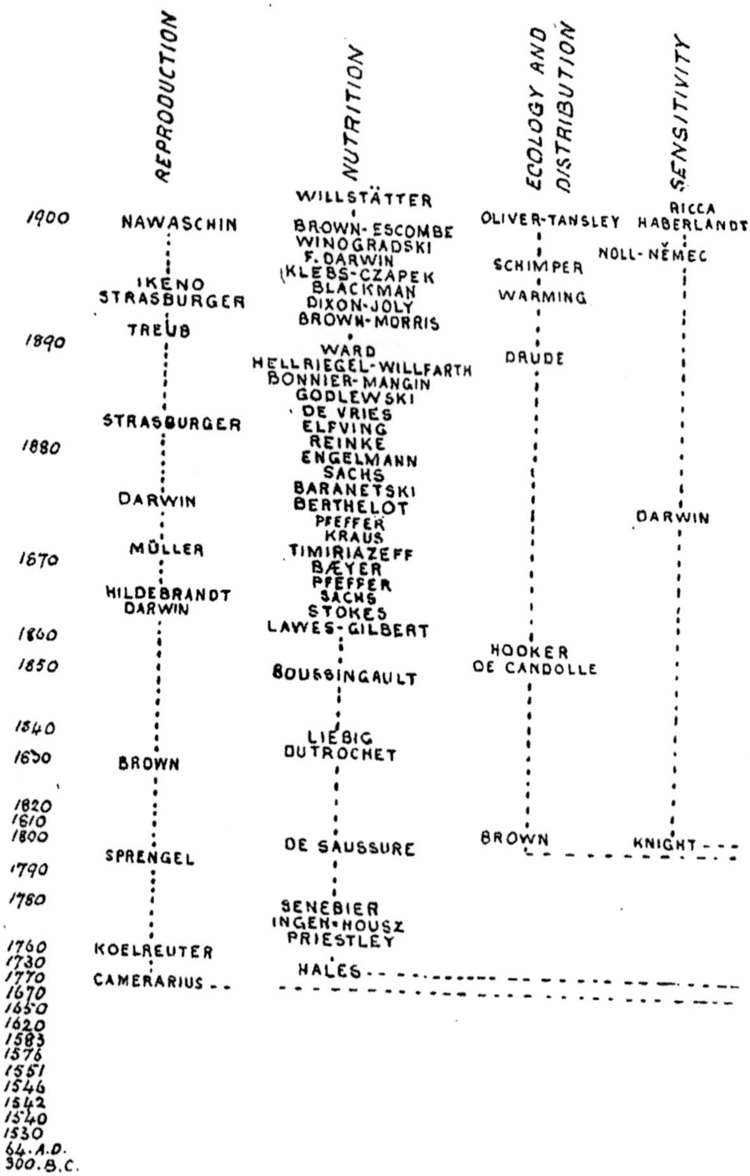
REPRODUCTION
NUTRITION
ECOLOGY AND DISTRIBUTION
SENSITIVITY
1900
1890
1880
1870
1860
1850
1840
1830
1820
1810
1800
1790
1780
1760
1730
1770
1670
1650
1620
1583
1576
1551
1546
1542
1540
1530
64. A.D.
300. B.C.
NAWASCHIN
IKENO
STRASBURGER
TREUB
STRASBURGER
DARWIN
MÜLLER
HILDEBRANDT
DARWIN
BROWN
SPRENGEL
KOELREUTER
CAMERARIUS
WILLSTÄTTER
BROWN-ESCOMBE
WINOGRADSKI
F. DARWIN
KLEBS-CZAPEK
BLACKMAN
DIXON-JOLY
BROWN-MORRIS
WARD
HELLRIEGEL-WILLFARTH
BONNIER-MANGIN
GODLEWSKI
DE VRIES
ELFVING
REINKE
ENGELMANN
SACHS
BARANETSKI
BERTHELOT
PFEFFER
KRAUS
TIMIRIAZEFF
BÆYER
PFEFFER
SACHS
STOKES
LAWES-GILBERT
BOUSSINGAULT
LIEBIG
DUTROCHET
DE SAUSSURE
SENEBIER
INGEN-HOUSZ
PRIESTLEY
HALES
OLIVER-TANSLEY
SCHIMPER
WARMING
DRUDE
HOOKER
DE CANDOLLE
BROWN
RICCA
HABERLANDT
NOLL-NĚMEC
DARWIN
KNIGHT

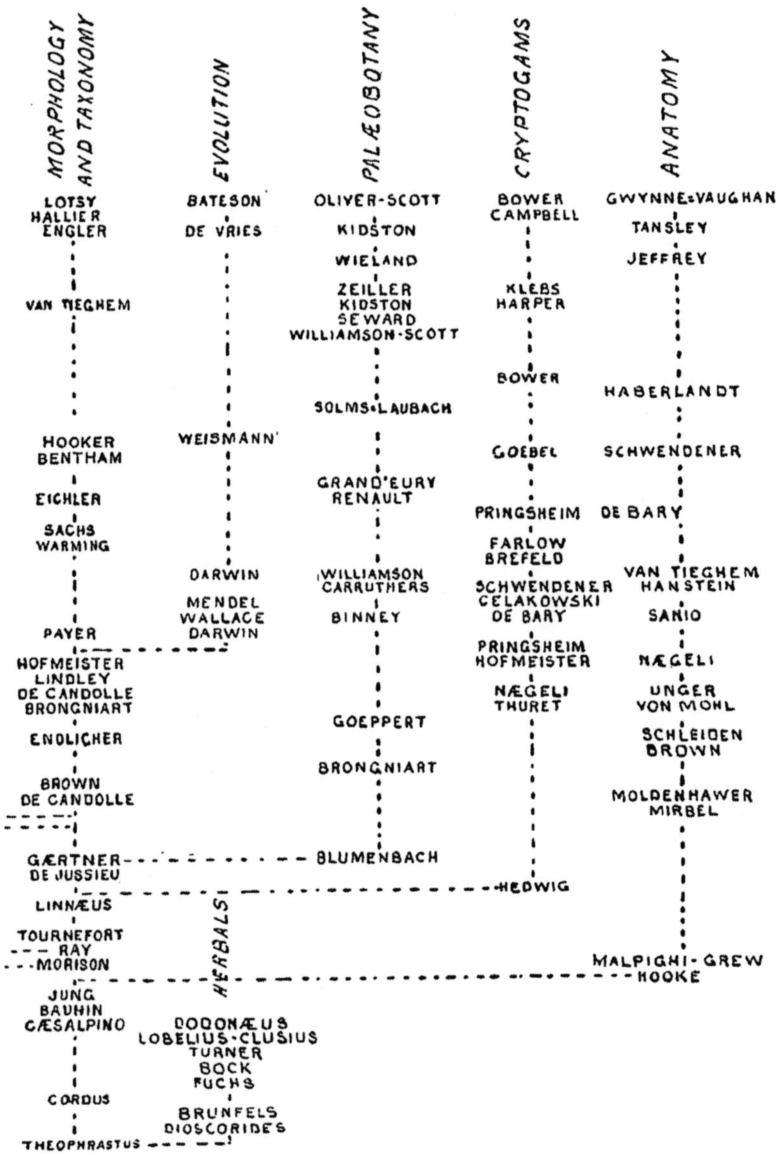
MORPHOLOGY AND TAXONOMY
EVOLUTION
PALÆOBOTANY
CRYPTOGAMS
ANATOMY
LOTSY
HALLIER
ENGLER
VAN TIEGHEM
HOOKER
BENTHAM
EICHLER
SACHS
WARMING
PAYER
HOFMEISTER
LINDLEY
DE CANDOLLE
BRONGNIART
ENDLICHER
BROWN
DE CANDOLLE
GÆRTNER
DE JUSSIEU
LINNÆUS
TOURNEFORT
RAY
MORISON
JUNG
BAUHIN
CÆSALPINO
CORDUS
THEOPHRASTUS
BATESON
DE VRIES
WEISMANN
DARWIN
MENDEL
WALLACE
DARWIN
HERBALS
DODONÆUS
LOBELIUS-CLUSIUS
TURNER
BOCK
FUCHS
BRUNFELS
DIOSCORIDES
OLIVER-SCOTT
KIDSTON
WIELAND
ZEILLER
KIDSTON
SEWARD
WILLIAMSON-SCOTT
SOLMS-LAUBACH
GRAND'EURY
RENAULT
WILLIAMSON
CARRUTHERS
BINNEY
GOEPPERT
BRONGNIART
BLUMENBACH
BOWER
CAMPBELL
KLEBS
HARPER
BOWER
GOEBEL
PRINGSHEIM
FARLOW
BREFELD
SCHWENDENER
CELAKOWSKI
DE BARY
PRINGSHEIM
HOFMEISTER
NÆGELI
THURET
HEDWIG
GWYNNE-VAUGHAN
TANSLEY
JEFFREY
HABERLANDT
SCHWENDENER
DE BARY
VAN TIEGHEM
HANSTEIN
SANIO
NÆGELI
UNGER
VON MOHL
SCHLEIDEN
BROWN
MOLDENHAWER
MIRBEL
MALPIGHI-GREW
HOOKE

INDEX

Printed in Great Britain by R. & R. Clark, Limited, *Edinburgh.*